普通高等教育"十三五"规划教材

能源与动力工程专业英语

《能源与动力工程专业英语》编写组　编

中国石化出版社

内 容 提 要

本书由精通英语的能源与动力工程专业的教师在总结多年科技英语和专业英语教学经验的基础上编写而成。全书包括热工基础、热工设备、热工应用三个部分。其中热工基础部分包括传热与热力学、燃料与燃烧、热工仪表及控制原理等内容；热工设备部分包括锅炉、汽轮机、制冷设备等内容；热工应用部分包括热能利用及专业英语写作等内容。全书共有44个单元，每个单元则由课文、课文词汇表、练习作业等组成。

本书有较强的知识性和实用性，可作为高等院校能源动力类本科生的专业英语教材，也适合作为从事相关专业的工程技术人员的学习参考材料。

图书在版编目(CIP)数据

能源与动力工程专业英语/《能源与动力工程专业英语》编写组编.—2版.—北京:中国石化出版社,2018.8(2020.8重印)
普通高等教育“十三五”规划教材
ISBN 978-7-5114-4975-7

Ⅰ.①能… Ⅱ.①能… Ⅲ.①能源-英语-高等学校-教材②动力工程-英语-高等学校-教材 Ⅳ.①TK

中国版本图书馆CIP数据核字(2018)第182271号

中国石化出版社出版发行
地址:北京市东城区安定门外大街58号
邮编:100011 电话:(010)57512500
发行部电话:(010)57512575
http://www.sinopec-press.com
E-mail:press@sinopec.com
北京柏力行彩印有限公司印刷
全国各地新华书店经销
*
787×1092毫米 16开本 9.5印张 235千字
2018年8月第2版 2020年8月第2次印刷
定价:30.00元

第二版前言

本书第一版《热能与动力工程专业英语》于2008年出版，出版后得到许多院校师生的认可，多次重印。本版根据高等院校新的专业名称设置，更名为《能源与动力工程专业英语》。

大学本科学生在学完基础英语之后，在高年级可通过学习专业英语课程做到大学四年英语学习不间断，熟悉科技与专业英语的特点，扩大科技与专业英语词汇量，提高科技与专业英语阅读能力。本教材是在目前尚不具备在某些专业课中直接使用英文教材和使用英语讲授的条件下，由精通英语的能源与动力工程专业的专业教师在总结多年科技英语和专业英语教学经验的基础上编写成的。通过学习本书，可使能源与动力工程专业的大学本科生巩固专业知识，提高专业英语的阅读水平。

本书覆盖了能源与动力工程专业的基本内容。全书包括传热与热力学、燃料与燃烧、热工仪表及控制原理、锅炉、汽轮机、制冷设备、换热器、能源利用及专业英语写作等内容。

本书选材较为新颖，文体规范，难度适中。为了适应专业英语教学方面的要求，本书既全面覆盖了学生学过的内容，又拓宽了专业领域的知识。书中每篇英文阅读材料都安排了课后练习，可使学生对所学知识进一步巩固与提高。为了便于读者阅读本书，在每篇阅读材料后都附有词汇表。为提高学生的专业英语综合能力，本书还特地增加了专业英语写作方面的内容。

全书共分三个部分，其中第一部分由战洪仁、王立鹏、寇丽萍、王翠华编写。第二部分由王翠华、张先珍、李雅侠、战洪仁、曾祥福编写。第三部分由曾祥福、张先珍、寇丽萍、王立鹏、李雅侠编写。本书由沈阳化工大学金志浩教授审阅。感谢刘鹏、刘彦超参与本书的校对。

由于编者水平有限，书中不足之处在所难免，恳请广大读者批评指正。

编　者

目 录

PART Ⅰ BASIC OF PYROLOGY

Unit 1 Heat Transfer

Heat transfer is the science that seeks to predict the energy transfer that may take place between material bodies as a result of a temperature difference.

The usual progression in the educational process of the thermal systems engineering generally starts with the study of energy in a beginning physics course, progresses to the thermodynamics sequence, followed by a beginning heat transfer course, and then goes on to specialized courses in more advanced treatment of the basic modes of heat transfer and/or to applications oriented courses.

Thermodynamics teaches that this energy transfer is defined as heat. The science of heat transfer seeks not merely to explain how heat energy may be transferred, but also to predict the rate at which the exchange will take place under certain specified conditions. The fact that a heat-transfer rate is the desired objective of an analysis points out the difference between heat transfer and thermodynamics. Thermodynamics deals with systems in equilibrium; it may be used to predict the amount of energy required to change a system from one equilibrium state to another; it may not be used to predict how fast a change will take place since the system is not in equilibrium during the process. Heat transfer supplements the first and second principles of thermodynamics by providing additional experimental rules which may be used to establish energy-transfer rates. As in the science of thermodynamics, the experimental rules used as a basis of the subject of heat transfer are rather simple and easily expanded to encompass a variety of practical situations.

As an example of the different kinds of problems that are treated by thermodynamics and heat transfer, consider the cooling of a hot steel bar that is placed in a pail of water. Thermodynamics may be used to predict the final equilibrium temperature of the steel bar-water combination. Thermodynamics will not tell us how long it takes to reach this equilibrium condition or what the temperature of the bar will be after a certain length of time before the equilibrium condition is attained. Heat transfer may be used to predict the temperature of both the bar and the water as a function of time.

One generally identifies three basic mechanisms of transport: conduction, convection, and radiation. Of these mechanisms, conduction and radiation can be considered as pure in the sense that they can take place as the only propagating mechanisms. Convection, on the other hand, is a mixture of conduction and mass transport of energy, with radiation present in significant or insignificant amounts depending on the fluid present and the temperature levels. One might also want to include phase change as a basic mechanism; however, even more so than convection, phase change is a mixture of conduction and complicate mass transport processes in the fluid portion, in addition the actual change of phase mechanism. It is, therefore, generally considered to be in a category by itself. Since most heat transfer occurrences involve more than one mode, it will be necessary to critically examine the basic modes before considering more general happenings. In the discussion to follow, then, we focus only on the three basic modes and their concepts and definitions.

Conduction Heat Transfer

When a temperature gradient exists in a body, experience has shown that there is an energy transfer from the high-temperature region to the low-temperature region. We say that the energy is transferred by conduction and that the heat-transfer rate per unit area is proportional to the normal temperature gradient:

$$\frac{q}{A} \sim \frac{\partial T}{\partial x}$$

When the proportionality constant is inserted,

$$Q = -kA\frac{\partial T}{\partial x} \tag{1-1}$$

Where Q is the heat-transfer rate and $\partial T/\partial x$ is the temperature gradient in the direction of the heat flow in calculus notation. When the temperature is a function of only one variable (distance x, for example), we can express the temperature gradient by means of the differential operator d as $\mathrm{d}T/\mathrm{d}x$. We can then express Equation (1-2) in the following form:

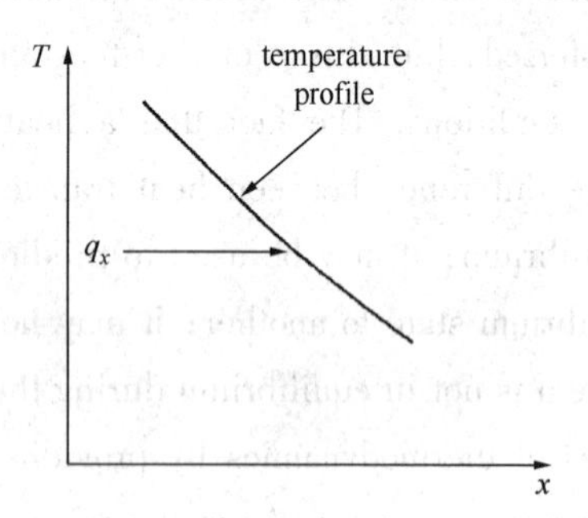

Fig.1.1 Sketch showing direction of heat flow

$$Q = -kA\frac{\mathrm{d}T}{\mathrm{d}x} \tag{1-2}$$

The positive constant k is called the thermal conductivity of the material, and the minus sign is inserted so that the second principle of thermodynamics will be satisfied; i.e., heat must flow downhill on the temperature scale, as indicated in the coordinate system of Fig 1.1. Equation (1-1) is called Fourier's law of heat conduction after the French mathematical physicist Joseph Fourier, who made very significant contributions to the analytical treatment of conduction heat transfer. It is important to note that Equation (1-1) is the defining equation for the thermal conductivity and that k has the units of watts per meter per Celsius degree in a typical system of units in which the heat flow is expressed in watts.

The thermal-conductivity k is generally a function of temperature. For solids, the variation with temperature is usually not a strong one, but for liquids and gases, k can be a strong function of temperature.

Words and Expressions

1. thermodynamics [θəːməudaiˈnæmiks] *n.* 热力学
2. combination [ˌkɔmbiˈneiˈʃən] *n.* 结合，联合，合并，化合，化合物
3. function [ˈfʌŋkʃən] *n.* 功能，作用；函数
4. conduction [kənˈdʌkʃən] *n.* 引流；输送，传播；传导性[率]
5. convection [kənˈvekʃən] *n.* (=transmission) 传达；(热) 对流

6. radiation [ˌreidiˈeiʃən] *n.* 放射，辐射;发射;发光
7. mechanisms [ˈmekənizəm] *n.* 机械，机械装置[结构]
8. significant [sigˈnifikənt] *adj.* 有意义的，重大的，重要的;表明……的 (of)
9. gradient [ˈgreidiənt] *n.* 坡度;梯度;陡度;斜率;（温度，气压的）变化率
10. proportionality [prəˌpɔːʃəˈnæliti] *n.* 比例（性），均衡（性），相称
11. calculus [ˈkælkjuləs] *n.* 微积分(学)
12. heat transfer 传热（学）
13. starts with 以……开始
14. principles of thermodynamics 热力学定律
15. energy-transfer 能量转换
16. equilibrium temperature 平衡温度
17. on the other hand 另一方面
18. phase change 相变，换相
19. in addition 另外
20. temperature gradient 温度梯度
21. proportionality constant 比例常数
22. thermal conductivity 导热性（系数）

Exercises

1. Put the following into Chinese.

(1) energy-transfer (2) equilibrium temperature (3) principles of thermodynamics
(4) temperature gradient (5) heat transfer

2. Answer the following question, according to text.

(1) What is heat transfer?
(2) What is the difference between heat transfer and thermodynamics?

3. Translate the paragraph 3, 4 of the text into Chinese.

Unit 2 The Convection Mode

When a fluid at rest or in motion is in contact with a surface at a temperature different from the plate, energy flows in the direction of the lower temperature as required by the principle of thermodynamics. We say that heat is convected away, and we call the process convection heat transfer. Fig. 1.2 shows some possibilities.

For both situations shown in Fig. 1.2, we express the overall effect of convection, we use Newton's law of cooling:

$$Q = hA(T_W - T_\infty) \tag{1-3}$$

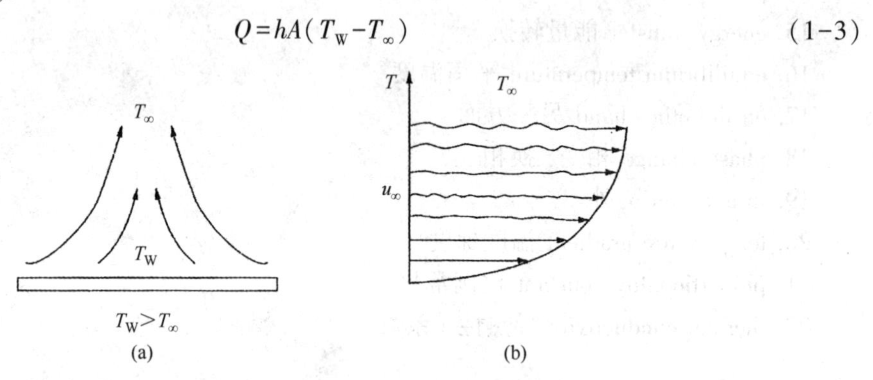

Fig. 1.2 Convection concepts for a fluid in contact with a solid surface:
(a) fluid far from surface at rest, and (b) fluid far from surface. In motion with respect to the surface.

The temperature T_W is that directly at the surface in contact with the plate, and the temperature T_∞ is the fluid temperature far enough away from the surface so that no influence of the surface is evident. The area A is the surface area in contact with the fluid and we should note that A is perpendicular to the direction of the heat flux Q. The proportionality factor h is called the heat transfer coefficient (also the unit area conductance or the convective conductance) and depends on the geometrical arrangement, orientation, and surface condition (smooth or rough), as well as on the properties and velocity of the fluid.

There are two convection modes: forced convection and natural convection. If a heated plate were exposed to ambient room air without an external source of motion, a movement of the air would be experienced as a result of the density gradients near the plate. We call this natural convetion (or free convection), as opposed to forced convection, which is experienced in the case of the fan blowing air over a plate.

Table 1.1 lists some representative values for the heat transfer coefficient under a variety of engineering conditions.

Table 1.1 Representative values for the heat transfer coefficient

Condition	h/ (Btu/h · ft^2 · F)	h/ (kW/m^2 · K)
Free convection, air	1~6	0.006~0.035

续表

Condition	h/ (Btu/h · ft² · F)	h/ (kW/m² · K)
Forced convection, air	5~150	0.028~0.851
Free convection, water	30~200	0.170~1.14
Forced convection, water	100~4000	0.570~22.7
Boiling water	1000~15000	5.70~85
Condensing steam	10000~30000	57~170
Forced convection, sodium	20000~40000	113~227

Convection Energy Balance on a Flow Channel

The energy transfer expressed by Equation (1-3) is used for evaluating the convection loss for flow over an external surface. Of equal importance is the convection gain or loss resulting from a fluid flowing inside a channel or tube, as shown in Fig. 1.3. In this case, the heated wall at T_W loses heat to the cooler fluid which consequently rises in temperature as it flows from inlet conditions at T_i to exit conditions at T_e. Using the symbol i to designate enthalpy (to avoid confusion with h, the convection coefficient), the energy balance on the fluid is

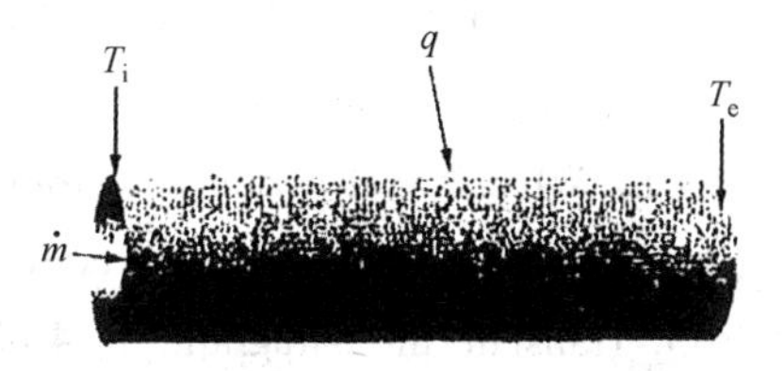

Fig. 1.3 Convection in a channel

$$q=\dot{m}(i_e-i_i) \quad (1-4)$$

Where $\dot{m}$ is the fluid mass flow rate. For many single phase liquids and gases operating over reasonable temperature ranges $\Delta i=c_p\Delta T$ and we have

$$q=\dot{m}c_p(T_e-T_i)$$

which may be equated to a convection relation like Equation (1-3)

$$q=\dot{m}c_p(T_e-T_i)=hA(T_{w,\ avg}-T_{fluid,\ avg}) \quad (1-5)$$

In this case, the fluid temperatures T_e, T_i and T_{fluid} are called bulk or energy average temperatures. A is the surface area of the flow channel in contact with the fluid.

Words and Expressions

1. perpendicular [ˌpəːpənˈdikjulə] *adj.* 垂直的，正交的；*n.* 垂线
2. property [ˈprɔpəti] *n.* 性质，特性
3. representative [ˌrepriˈzentətiv] *n.* 代表；*adj.* 典型的，有代表性的
4. consequently [kɔnsikwəntli] *adv.* 从而，因此
5. confusion [kənˈfjuːʒən] *n.* 混乱，混淆
6. enthalpy [ˈenθælpi, enˈθælpi] *n.*【物】焓，热函
7. at rest 安眠，长眠，静止
8. surface area 表面积
9. convection heat transfer 对流传热
10. Newton's law of cooling 牛顿冷却定律

11. proportionality factor 比例因子，比例系数
12. heat transfer coefficient 传热系数
13. properties of the fluid 流体的物性
14. forced convection 强迫对流
15. natural convection 自然对流
16. flow channel 液流通路;流动（试验）水槽;气流道
17. mass flow rate 质量流量
18. average temperatures 平均温度

Exercises

1. Put the following into Chinese.

(1) forced convection　(2) natural convection　(3) average temperatures

(4) at rest　(5) mass flow rate

2. Answer the following question, according to text.

(1) What is convection heat transfer?

(2) What are the modes of convection heat transfer?

3. Translate the paragraph 3, 4 of the text into Chinese.

Unit 3 Radiation Heat Transfer

In contrast to the mechanisms of conduction and convection, where energy transfer through a material medium is involved, heat may also be transferred through regions where a perfect vacuum exists. The mechanism in this case is electromagnetic radiation. We shall limit our discussion to e-lectromagnetic radiation which is propagated as a result of a temperature difference; this is called thermal radiation.

Thermodynamic considerations show that an ideal thermal radiator, or blackbody, will emit energy at a rate proportional to the fourth power of the absolute temperature of the body and directly proportional to its surface area. Thus

$$q_{emittid} = \sigma A T^4 \qquad (1-6)$$

Where σ is the proportionality constant and is called the Stefan-Boltzmann constant with the value of $5.669 \times 10^{-8} W/m^2 \cdot K^4$. Equation (1-6) is called the Stefan-Boltzmann law of thermal radiation, and it applies only to blackbodies. It is important to note that this equation is valid only for thermal radiation; other types of electromagnetic radiation may not be treated so simply.

Equation (1-6) governs only radiation emitted by a blackbody. The net radiant exchange between two surfaces will be proportional to the difference in absolute temperatures to the fourth power, i. e.

$$\frac{q_{netexchange}}{A} \propto \sigma \left(T_1^4 - T_2^4\right) \qquad (1-7)$$

Where $q_{netexchange}$ is the net radiant exchange.

We have mentioned that a blackbody is a body that radiates energy according to the T^4 law. We call such a body black because black surfaces, such as a piece of metal covered with carbon black, approximate this type of behavior other types of surfaces, such as a glossy painted surface or a polished metal plate, do not radiate as much energy as the blackbody; however, the total radiation emitted by these bodies still generally follows the ${T_1}^4$ proportionality. To take account of the "gray" nature of such surfaces, we introduce another factor into Equation (1-7), called the emissive $\in$, which relates the radiation of the "gray" surface to that of an ideal black surface. In addition, we must take into account the fact that not all the radiation leaving one surface will reach the other surface since electromagnetic radiation travels in straight lines and some will be lost to the surroundings. We therefore introduce two new factors in Equation (1-6) to take into account both situations, so that

$$q = F_{\in} F_G \sigma A \left(T_1^4 - T_2^4\right) \qquad (1-8)$$

where $F_{\in}$ is an emissive function, and F_G is a geometric "view factor" function.

Radiation in an Enclosure

A simple radiation problem is encountered when we have a heat-transfer surface at temperature T_1 completely enclosed by a much larger surface maintained at T_2. The net radiant exchange in this case can be calculated with

$$q = \in_1 \sigma A_1 (T_1^4 - T_2^4) \qquad (1-9)$$

Values of ∈ are given in Appendix A.

Radiation heat-transfer phenomena can be exceedingly complex, and the calculations are seldom as simple as implied by Equation (1-9).

Summary

We may summarize our introductory remarks very simply. Heat transfer may take place by one or more of three modes: conduction, convection, and radiation. It has been noted that the physical mechanism of convection is related to the heat conduction through the thin layer of fluid adjacent to the heat-transfer surface. In both conduction and convection Fourier's law is applicable, although fluid mechanics must be brought into the convection problem in order to establish the temperature gradient.

Radiation heat transfer involves a different physical mechanism-that of propagation of electromagnetic energy. To study this type of energy transfer we introduce the concept of an ideal radiator or blackbody, which radiates energy at a rate proportional to its absolute temperature to fourth power.

It is easy to envision cases in which all three modes of heat transfer are present, as in Fig. 1.4. In this case the heat conducted through the plate is removed from the plate surface by a combination of convection and radiation. An energy balance was given as

$$-kA \left.\frac{dT}{dy}\right|_{wall} = hA\ (T_W - T_\infty)\ + F_\in F_G \sigma A\ (T_W^4 - T_S^4) \qquad (1-10)$$

where

T_S = temperature of surroundings

T_W = surface temperature

T_∞ = fluid temperature

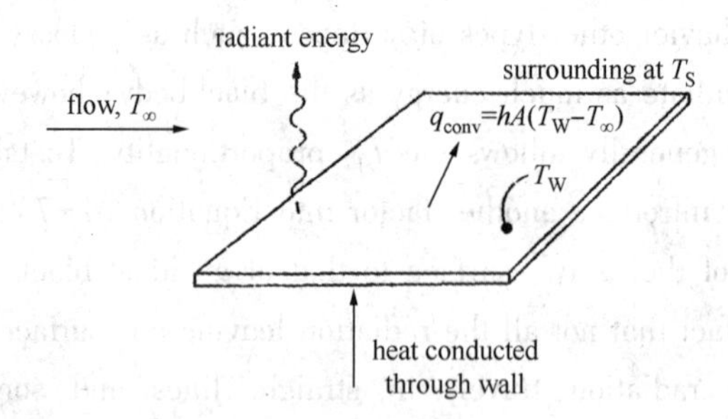

Fig. 1.4 Combination of conduction, convection, and radiation heat transfer

To apply the science of heat transfer to practical situations, a thorough knowledge of all three modes of heat transfer must be obtained.

Words and Expressions

1. electromagnetic [ilektrəʊ'mægnetik] *adj.* 电磁的
2. Stefan-Boltzmann 斯蒂藩波尔兹曼

3. radiate ['reidieit] *vt.* 放射，辐射，传播，广播；*vi.* 发光，辐射，流露
4. emissive [i'misiv] *n.* 【物】发射率
5. geometric [dʒiə'metrik] *adj.* 几何的，几何学的
6. enclosure [in'kləuʒə] *n.* 围住，围栏，四周有篱笆或围墙的场
7. in contrast to 和……形成对比
8. electromagnetic radiation 电磁辐射
9. thermal radiation 热辐射
10. thermal reactor 热反应堆
11. Stefan-Boltzmann law 斯蒂藩波尔兹曼定律
12. carbon black 黑烟末
13. in addition 另外
14. take into account 重视，考虑
15. view factor 角系数
16. blackbody 黑体

Exercises

1. Put the following into Chinese.

(1) in contrast to (2) carbon black (3) in addition

(4) take into account (5) Stefan-Boltzmann law

2. Answer the following question, according to text.

(1) What is radiation heat transfer?

(2) What are the factor related to radiation heat transfer?

3. Translate the paragraph 1, 4 of the text into Chinese.

Unit 4 Basic Concepts of Thermodynamics

Thermodynamics is a basic science that deals with energy and has long been an essential part of engineering curricula all over the world. This introductory text contains sufficient material for two sequential courses in thermodynamics, and it is intended for use by undergraduate engineering students and by practicing engineers as a reference. The objectives of this text are to cover the basic principle of thermodynamics.

The basic abstraction of thermodynamics is the division of the world into systems delimited by real or ideal boundaries. The systems not directly under consideration are lumped into the environment. It is possible to subdivide a system into subsystems, or to group several systems together into a larger system. Usually systems can be assigned a well-defined state which can be summarized by a small number of parameters.

Thermodynamic Systems

A thermodynamic system is that part of the universe that is under consideration. A real or imaginary boundary separates the system from the rest of the universe, which is referred to as the environment. A useful classification of thermodynamic systems is based on the nature of the boundary and the flows of matter, energy and entropy through it. There are three kinds of systems depending on the kinds of exchanges taking place between a system and its environment:

- *isolated* systems: not exchanging heat, matter or work with their environment. An example of an isolated system would be an insulated container, such as an insulated gas cylinder.
- *closed* systems: exchanging energy (heat and work) but not matter with their environment. A greenhouse is an example of a closed system exchanging heat but not work with its environment. Whether a system exchanges heat, work or both is usually thought of as a property of its boundary, which can be

  * *adiabatic* boundary: not allowing heat exchange;
  * *rigid* boundary: not allowing exchange of work.

- *open* systems: exchanging energy (heat and work) and matter with their environment. A boundary allowing matter exchange is called permeable. The ocean would be an example of an open system.

In reality, a system can never be absolutely isolated from its environment, because there is always at least some slight coupling, even if only via minimal gravitational attraction. In analyzing an open system, the energy into the system is equal to the energy leaving the system.

Thermo Dynamics and Energy

Thermodynamics can be defined as the science of energy. Although everybody has a feeling of what energy is, it is difficult to give a precise definition for it. Energy can be viewed as the ability to cause changes.

The name thermodynamics stems from the Greek words therme (heat) and dynamics (power). which is most descriptive of the early efforts to convert heat into power. Today the same name is broadly interpreted to include all aspects of energy and energy transformations, including power generation, refrigeration and relationships among the properties of matter.

One of the most fundamental laws of nature is the conservation of energy principle. It simply states that during an interaction, energy can change from one form to another but the total amount of energy remains constant. That is, energy cannot be created or destroyed. A rock falling off a cliff, for example, picks up speed as a result of its potential energy being converted to kinetic energy (Fig. 1.5). The conservation of energy principle also forms the backbone of the diet industry: a person who has a greater energy input (food) than energy output (exercise) will gain weight (store energy in the form of fat), and a person who has a smaller energy input than output will lose weight (Fig. 1.6). The change in the energy content of a body or any other system is equal to the difference between the energy input and the energy output, and the energy balance is expressed as $E_{in} - E_{out} = \Delta E$.

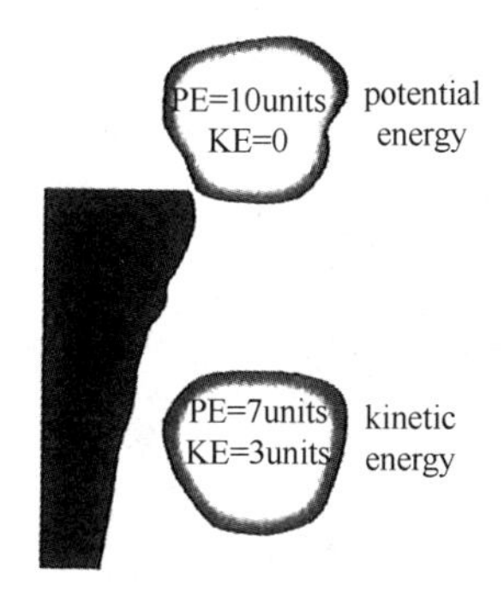

Fig.1.5 Energy cannot be created or destroyed; It can only change forms (the first law)

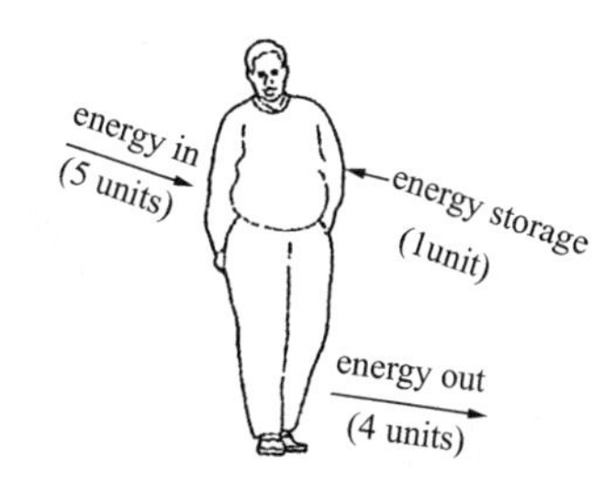

Fig.1.6 Conservation of energy principle for the human body

Words and Expressions

1. boundary [ˈbaundəri] *n.* 边界，控制面
2. entropy [ˈentrəpi] *n.* [物]熵，[无]平均信息量
3. permeable [pəːmiəbl] *adj.* 有浸透性的，能透过的
4. fundamental [ˌfʌndəˈmentl] *adj.* 基础的，基本的；*n.*基本原则，基本原理
5. absolutely [ˈæbsəluːtli] *adv.* 完全地，绝对地
6. kinetic [kaiˈnetik] *adj.* (运) 动的，动力 (学) 的
7. dynamics [daiˈnæmiks] *n.* 动力学
8. conservation [ˌkɔnsə(ː)ˈveiʃən] *n.* 保存，保持，守恒
9. energy principle 能量原理
10. conservation of energy 能量守恒
11. power generation 发电
12. energy balance 能量平衡
13. kinetic energy 动能
14. isolated systems 孤立系统

15. closed systems 封闭系统

16. open system 开口系统

Exercises

1. Put the following into Chinese.

(1) kinetic energy (2) power generation (3) isolated systems

(4) closed systems (5) energy balance

2. Answer the following question, according to text.

(1) What is the isolated systems?

(2) What is the open systems?

3. Translate the paragraph 1, 3 of the text into Chinese.

Unit 5 Laws of Thermodynamics

In simplest terms, the Laws of Thermodynamics dictate the specifics for the movement of heat and work. Basically, the First Law of Thermodynamics is a statement of the conservation of energy-the Second Law is a statement about the direction of that conservation-and the Third Law is a statement about reaching absolute Zero (0K).

First Law of Thermodynamics

The first law of thermodynamics is a statement of the principle of conservation of energy. It can also be considered as defining a property, the internal energy. The internal energy is all the energy associated with a substance, excluding potential and kinetic energies. It is a function of the thermodynamic state of the substance. Recall that both work and heat transfer are path functions. Although their values depend on the path chosen between two states, in the absence of changes in kinetic and potential energies, the difference between them has a unique value, whatever be the path chosen. As the difference between the heat and work transfer rates for any process between the two states has a unique value, the difference is related to a property in each state; the property is the internal energy. Thus in the absence of changes in kinetic and potential energies, for a control mass of a simple substance that undergoes a process from state 1 to state 2.

$$Q_2 = U_2 - U_1 + {}_1W_2$$

Where

Q_2 = heat transfer to the substance from the surroundings

U_2, U_1 = internal energies at states 2 and 1

${}_1W_2$ = work transfer from the substance to the surroundings

The Second Law of Thermodynamics

The Second Law of Thermodynamics states that' in all energy exchanges, if no energy enters or leaves the system, the potential energy of the state will always be less than that of the initial state'. This is also commonly referred to as entropy. A watchspring-driven watch will run until the potential energy in the spring is converted, and not again until energy is reapplied to the spring to rewind it. A car that has run out of gas will not run again until you walk 10 miles to a gas station and refuel the car. Once the potential energy locked in carbohydrates is converted into kinetic energy (energy in use or motion), the organism will get no more until energy is input again. In the process of energy transfer, some energy will dissipate as heat. Entropy is a measure of disorder: cells are not disordered and so have low entropy. The flow of energy maintains order and life. Entropy wins when organisms cease to take in energy and die.

The Third Law of Thermodynamics

The third law of thermodynamics is usually stated as a definition: **the entropy of a perfect crystal of an element at the absolute zero of temperature is zero.**

At the absolute zero of temperature, there is zero thermal energy or heat. Since heat is a measure of average molecular motion, zero thermal energy means that the average atom does not move at all. Since no atom can have less than zero motion, the motion of every individual atom must be zero when the average molecular motion is zero. When none of the atoms which make up a perfectly ordered crystal move at all, there can be no disorder or different states possible for the crystal.

Taking the entropy of a perfect crystal of an element to be zero at the absolute zero of temperature establishes a method by which entropies of elements at any higher temperature can be determined. Since by definition $dS = q_{rev}/T$, the mathematical integral of dS from zero to any higher temperature T is the integral, over that temperature range, of q_{rev}/T. In other words, the difference $S - S_0$ is the integral from zero to a temperature T of $(C_p/T)\,dT$. The molar heat capacity at any temperature is a measurable quantity, and so this difference can be determined experimentally.

The third law of thermodynamics is simply the statement that **S_0 is zero by definition for a pure element**, and so if the heat capacity is measured under conditions of reversible heat flow, as it can be, and as a function of temperature at low temperatures, as it can be, then the entropy S of a pure element at any temperature T is given by:

S = the integral from zero to T of $(C_p/T)\,dT$

Application Areas of Thermodynamics

All activities in nature involve some interaction between energy and matter; thus, it is hard to imagine an area that does not relate to thermodynamics in some manner.

For example, a cup of hot coffee left on a table eventually cools, but a cup of cool coffee in the same room never gets hot by itself (Fig. 1.7). The high-temperature energy of the coffee is degraded (transformed into a less useful form at a lower temperature) once it is transferred to the surrounding air.

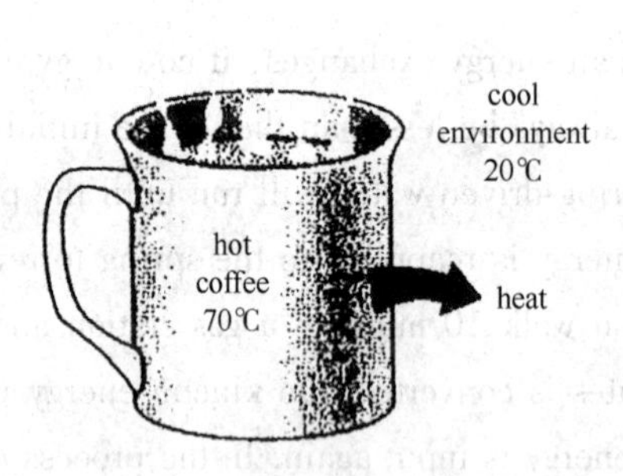

Fig.1.7 Heat flows in the direction of decreasing temperature

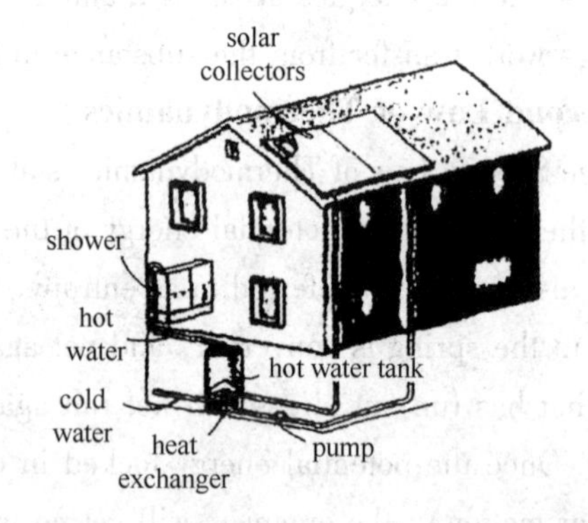

Fig.1.8 The design of many engineering systems, such as this solar hot water system, involves thermodynamics

An ordinary house is, in some respects, an exhibition hall filled with wonders of thermodynamics (Fig. 1.8). Many ordinary household utensils and appliances are designed, in whole or in part, by using the principles of thermodynamics. Some examples include the electric or gas range, the heating and air-conditioning systems, the refrigerator, the humidifier, the pressure cooker, the water heater, the shower, the iron, and even the computer, and the TV. On a larger scale, thermodynamics plays a major part in the design and analysis of automotive engines, rockets, jet engines, and conventional or nuclear power plants, solar collectors, and the design of vehicles from ordinary cars to airplanes (Fig. 1.9).

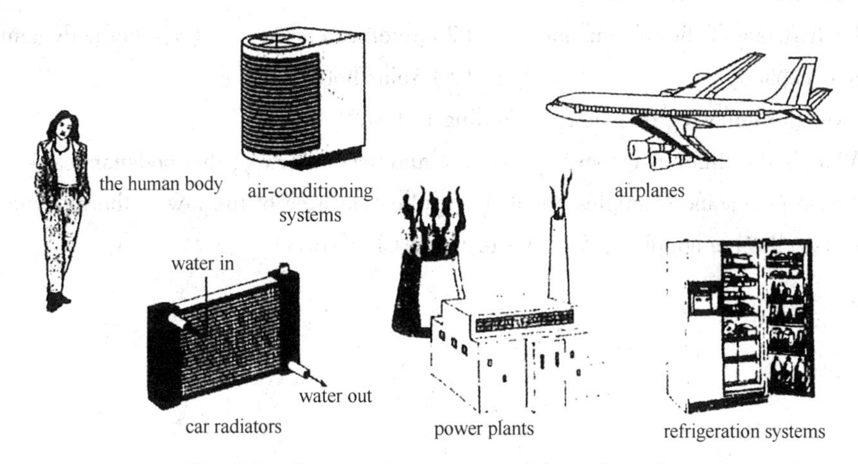

Fig. 1.9 Some application areas of thermodynamics

Words and Expressions

1. absolute [ˈæbsəluː t] *adj.* 完全的，绝对的
2. organism [ˈɔː gənizəm] *n.* 生物体，有机体
3. undergo [ˌʌndəˈgəu] *vt.* 经历，遭受，忍受
4. dissipate [ˈdisipeit] *v.* 驱散
5. disorder [disˈɔːdə] *n.* 杂乱，混乱，无秩序状态；*vt.* 扰乱，使失调，使紊乱
6. crystal [ˈkristl] *adj.* 结晶状的 *n.* 水晶，结晶，
7. mathematical [ˌmæθiˈmætikəl] *adj.* 数学的，精确的
8. integral[ˈintigrəl] *adj.* 完整的，【数学】积分的；*n.*【数学】积分，完整，部分
9. molar [ˈməulə] *adj.* 质量的，【化】【物】摩尔的
10. reversible [riˈvəːsəbl] *adj.* 可逆的
11. degraded [diˈgreidid] *adj.* 被降级的，退化的，
12. transformed [trænsˈfɔːm] *vt.* 转换，使……变形；*vi.* 转化，变换；*n.*【数】变换（式）
13. automotive [ɔːtəˈməutiv] *adj.* 汽车的，自动推进的
14. the first law of thermodynamics 热力学第一定律
15. internal energy 内能（热力学能）
16. thermodynamic state 热力学状态
17. molecular motion 分子运动
18. heat capacity 热容
19. jet engine 喷气机
20 solar collector 太阳能集热器

Exercises

1. Put the following into Chinese.

(1) the first law of thermodynamics (2) internal energy (3) thermodynamic state

(4) heat capacity (5) solar hot water

2. Answer the following question, according to text.

(1) What is the difference between the first and second law of thermodynamics?

(2) Please give some examples about the application area of the law of thermodynamics.

3. Translate the paragraph 4, 5 of the text into Chinese.

Unit 6 Thermodynamic state of a system

The thermodynamic state of a system is defined by specifying a set of measurable properties sufficient so that all remaining properties are determined. Examples of properties: pressure, temperature, density, internal energy, enthalpy, and entropy.

For engineering purposes we usually want gross, average, macroscopic properties (not what is happening to individual molecules and atoms) thus we consider substances as continua—the properties represent averages over small volumes. For example, there are 1016 molecules of air in 1 mm^3 at standard temperature and pressure.

Intensive properties do not depend on mass (e. g. p, T, ρ, $v=1/\rho$, u and h); extensive properties depend on the total mass of the system (e. g. V, M, U and H). Uppercase letters are usually used for extensive properties.

Equilibrium: States of a system are most conveniently described when the system is in equilibrium, i. e. it is in steady-state. Often we will consider processes that change "slowly"—termed quasi-equilibrium or quasi-static processes. A process is quasi-equilibrium if the time rate of change of the process is slow relative to the time it takes for the system to reach thermodynamic equilibrium. It is necessary that a system be quasi-equilibrium before applying many of the thermodynamics relations to that system.

This is illustrated in Fig. 1.10. When a gas in a piston-cylinder device is compressed suddenly, the molecules near the face of the piston will not have enough time to escape and they will have to pile up in a small region in front of the piston, thus creating a high-pressure region there. Because of this pressure difference, the system can no longer be said to be in equilibrium, and this makes the entire process non-quasi-equilibrium. However, if the piston is moved slowly, the molecules will have sufficient time to redistribute and there will not be a molecule pileup in front of the piston. As a result, the pressure inside the cylinder will always be uniform and will rise at the same rate at all locations. Since equilibrium is maintained at all times, this is a quasi-equilibrium process.

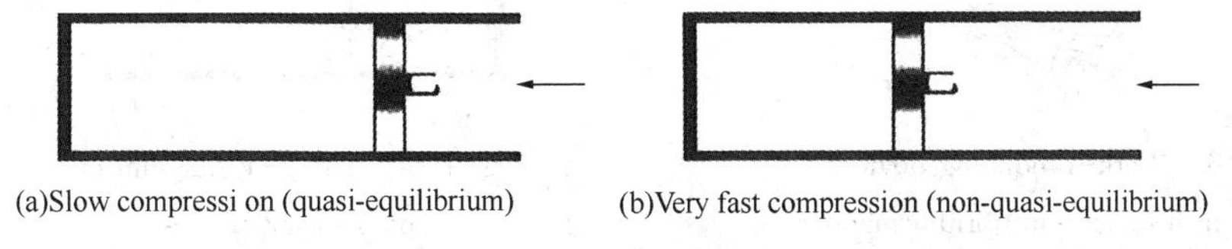

(a)Slow compressi on (quasi-equilibrium) (b)Very fast compression (non-quasi-equilibrium)

Fig. 1.10 Quasi-equilibrium and non-quasi-equilibrium compression

The state postulate requires that the two properties specified be independent to fix the state. Two properties are independent if one property can be varied while the other one is held constant. Temperature and specific volume, for example, are always independent properties, and together they can fix the state of a simple compressible system (Fig. 1.11).

Any change that a system undergoes from one equilibrium state to another is called a process, and the series of states through which a system passes during a process is called the path of the process (Fig. 1.12).

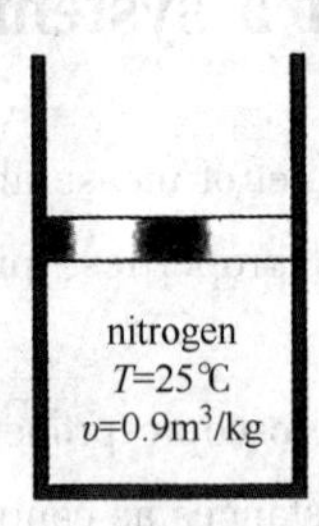

Fig.1.11 The state of nitrogen is fixed by two independent, intensive properties.

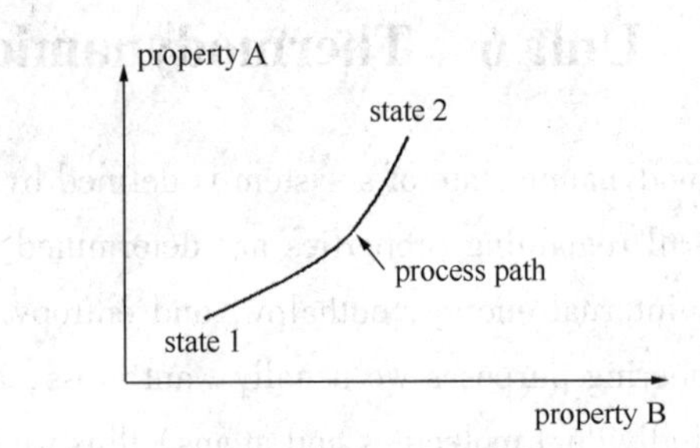

Fig.1.12 A process between slates 1 and 2 and the process path

It should be pointed out that a quasi-equilibrium process is an idealized process and is not a true representation of an actual process. But many actual processes closely approximate it, and they can be modeled as quasi-equilibrium with negligible error. Engineers are interested in quasi-equilibrium processes for two reasons, first, they are easy to analyze; second, work-producing devices deliver the most work when they operate on quasi-equilibrium processes (Fig. 1.13). Therefore, quasi-equilibrium processes serve as standards to which actual processes can be compared.

Process diagrams plotted by employing thermodynamic properties coordinates are very useful in visualizing the processes. Some common properties that are used as coordinates are temperature T, pressure p, and volume V (or specific volume v). Fig. 1.14 shows the p-V diagram of a compression process of a gas.

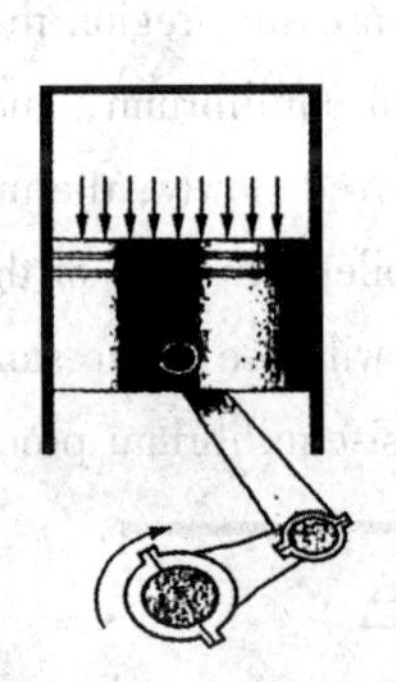

Fig. 1.13 Work-producing devices operating in a quasi-equilibrium manner deliver the most work

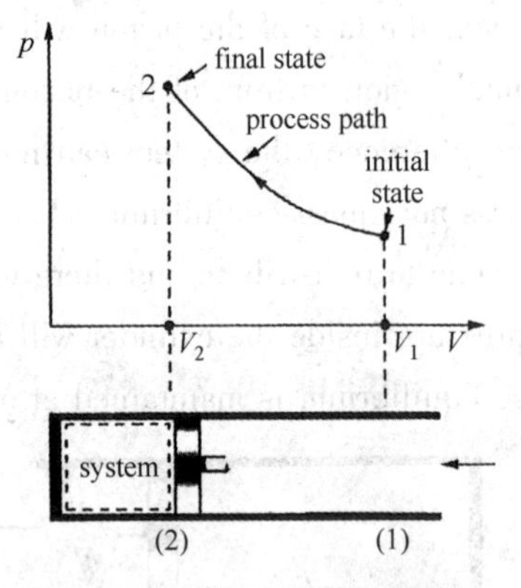

Fig. 1.14 The $p-V$ diagram of a compression process

Words and Expressions

1. property [ˈprɔpəti] *n.* 财产，所有物，所有权，性质，特性，(小) 道具
2. enthalpy [ˈenθælpi, enˈθælpi] *n.*【物】焓，热函
3. macroscopic [ˌmækrəuˈskɔpik] *adj.* 肉眼可见的，巨观的
4. standard [ˈstændəd] *n.* 标准，规格；*adj.* 标准的

5. uppercase ['ʌpə'keis] *n.* 大写字母盘
6. equilibrium [ˌiːkwi'libriəm] *n.* 平衡，均衡
7. escape [is'keip] *n.* 逃，逃亡，溢出设备；*vi.* 逃脱，避开，溜走；*vt.* 逃避，避免
8. independent [indi'pendənt] *n.* 中立派，无党派者；*adj.* 独立自主的，不受约束的
9. macroscopic property【核】宏观特性，【热】宏观性质
10. intensive property 强度性（质），强度参数
11. extensive property 广延性（质），广度参数
12. steady-state 不变的，永恒的
13. quasi-equilibrium 准平衡
14. thermodynamic equilibrium 热力平衡
15. specific votumen【力】比容
16. equilibrium state 平衡态
17. thermodynamic property 热力学性质
18. thermoelectric-ties【物】温差电性

Exercises

1. Put the following into Chinese.

(1) macroscopic property　(2) intensive property　(3) quasi-equilibrium
(4) specific volume　(5) thermodynamic property

2. Answer the following question, according to text.

(1) What is the intensive properties?
(2) What are conditions of quasi-equilibrium?

3. Translate the paragraph 4, 8 of the text into Chinese.

Unit 7 Fossil Fuels-Coal, Oil and Natural Gas

Where Fossil Fuels Come From

There are three major forms of fossil fuels: coal, oil and natural gas. All three were formed many hundreds of millions of years ago before the time of the dinosaurs-hence the name fossil fuels. The age they were formed is called the Carboniferous Period. It was part of the Paleozoic Era. "Carboniferous" gets its name from carbon, the basic element in coal and other fossil fuels.

The Carboniferous Period occurred from about 360 to 286 million years ago. At the time, the land was covered with swamps filled with huge trees, ferns and other large leafy plants, similar to the picture above. The water and seas were filled with algae—the green stuff that forms on a stagnant pool of water. Algae is actually millions of very small plants.

Some deposits of coal can be found during the time of the dinosaurs. For example, thin carbon layers can be found during the late Cretaceous Period (65 million years ago)—the time of Tyrannosaurus Rex. But the main deposits of fossil fuels are from the Carboniferous Period.

As the trees and plants died, they sank to the bottom of the swamps of oceans. They formed layers of a spongy material called peat. Over many hundreds of years, the peat was covered by sand and clay and other minerals, which turned into a type of rock called sedimentary.

More and more rock piled on top of more rock, and it weighed more and more. It began to press down on the peat. The peat was squeezed and squeezed until the water came out of it and it eventually, over millions of years, it turned into coal, oil or petroleum, and natural gas.

Coal

Coal is a hard, black colored rock-like substance. It is made up of carbon, hydrogen, oxygen, nitrogen and varying amounts of sulphur. There are three main types of coal-anthracite, bituminous and lignite. Anthracite coal is the hardest and has more carbon, which gives it a higher energy content. Lignite is the softest and is low in carbon but high in hydrogen and oxygen content. Bituminous is in between. Today, the precursor to coal-peat is still found in many countries and is also used as an energy source.

The earliest known use of coal was in China. Coal from the Fu-shun mine in northeastern China may have been used to smelt copper as early as 3, 000 years ago. The Chinese thought coal was a stone that could burn.

Coal is found in many of the lower 48 states of U. S. and throughout the rest of the world. Coal is mined out of the ground using various methods. Some coal mines are dug by sinking vertical or horizontal shafts deep under ground, and coal miners travel by elevators or trains deep under ground to dig the coal. Other coal is mined in strip mines where huge steam shovels strip away the top layers above the coal. The layers are then restored after the coal is taken away.

The coal is then shipped by train and boats and even in pipelines. In pipelines, the coal is ground up and mixed with water to make what's called a slurry. This is then pumped many miles through pipelines. At the other end, the coal is used to fuel power plants and other factories.

Oil or Petroleum

Oil is another fossil fuel. It was also formed more than 300 million years ago. Some scientists say that tiny diatoms are the source of oil. Diatoms are sea creatures the size of a pin head. They do one thing just like plants; they can convert sunlight directly into stored energy.

In the graphic on the Fig. 1.15, as the diatoms died they fell to the sea floor ①. Here they were buried under sediment and other rock ②. The rock squeezed the diatoms and the energy in their bodies could not escape. The carbon eventually turned into oil under great pressure and heat. As the earth changed and moved and folded, pockets where oil and natural gas can be found were formed ③.

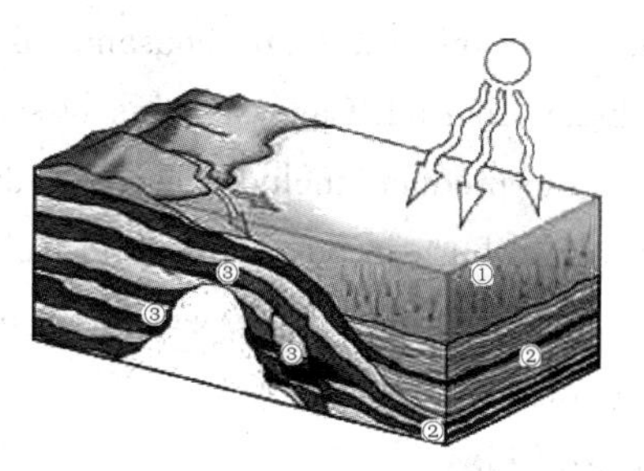

Fig. 1.15 The graphic of oil formation

Oil has been used for more than 5,000–6,000 years. The ancient Sumerians, Assyrians and Babylonians used crude oil and asphalt ("pitch") collected from large seeps at Tuttul (modern-day Hit) on the Euphrates River. A seep is a place on the ground where the oil leaks up from below ground. The ancient Egyptians, used liquid oil as a medicine for wounds, and oil has been used in lamps to provide light. The Dead Sea, near the modern Country of Israel, used to be called Lake Asphaltites. The word asphalt was derived is from that term because of the lumps of gooey petroleum that were washed up on the lake shores from underwater seeps.

In North America, Native Americans used blankets to skim oil off the surface of streams and lakes. They used oil as medicine and to make canoes water-proof. During the Revolutionary War, Native Americans taught George Washington's troops how to treat frostbite with oil.

As our country grew, the demand for oil continued to increase as a fuel for lamps. Petroleum oil began to replace whale oil in lamps because the price for whale oil was very high. During this time, most petroleum oil came from distilling coal into a liquid or by skimming it off of lakes-just as the Native Americans did.

Then on August 27, 1859, Edwin L. Drake (the man standing on the right in the black and white picture to the right), struck liquid oil at his well near Titusville, Pennsylvania. He found oil under ground and a way that could pump it to the surface. The well pumped the oil into barrels made out of wood. This method of drilling for oil is still being used today all over the world in areas where oil can be found below the surface.

Oil and natural gas are found under ground between folds of rock and in areas of rock that are porous and contain the oils within the rock itself. The folds of rock were formed as the earth shifts and moves. It's similar to how a small, throw carpet will bunch up in places on the floor.

To find oil and natural gas, companies drill through the earth to the deposits deep below the surface. The oil and natural gas are then pumped from below the ground by oil rigs (like in the picture). They then usually travel through pipelines or by ship.

Oil is brought to California by large tanker ships. The petroleum or crude oil must be changed or refined into other products before it can be used.

Oil is stored in large tanks until it is sent to various places to be used. At oil refineries, crude oil is split into various types of products by heating the thick black oil.

Oil is made into many different products-fertilizers for farms, the clothes you wear, the toothbrush you use, the plastic bottle that holds your milk, the plastic pen that you write with. They all came from oil. There are thousands of other products that come from oil. Almost all plastic comes originally from oil. Can you think of some other things made from oil?

The products include gasoline, diesel fuel, aviation or jet fuel, home heating oil, oil for ships and oil to burn in power plants to make electricity. Here's what a barrel of crude oil can make.

In California, 74 percent of our oil is used for transportation—cars, planes, trucks, buses and motorcycles.

Natural Gas

Sometime between 6,000 to 2,000 years BCE (Before the Common Era), the first discoveries of natural gas seeps were made in Iran. Many early writers described the natural petroleum seeps in the Middle East, especially in the Baku region of what is now Azerbaijan. The gas seeps, probably first ignited by lightning, provided the fuel for the "eternal fires" of the fire-worshiping religion of the ancient Persians.

Natural gas is lighter than air. Natural gas is mostly made up of a gas called methane. Methane is a simple chemical compound that is made up of carbon and hydrogen atoms. Its chemical formula is CH_4-one atom of carbon along with four atoms hydrogen. This gas is highly flammable.

Natural gas is usually found near petroleum underground. It is pumped from below ground and travels in pipelines to storage areas. The next chapter looks at that pipeline system.

Natural gas usually has no odor and you can't see it. Before it is sent to the pipelines and storage tanks, it is mixed with a chemical that gives a strong odor. The odor smells almost like rotten eggs. The odor makes it easy to smell if there is a leak.

Words and Expressions

1. pipeline [ˈpaipˌlain] *n.* 管道，输油管，输送管
2. sulphur [ˈsʌlfə] *n.*【化】硫（黄）
3. spongy [ˈspʌndʒi] *adj.* 像海绵的;海绵质的，多孔的，吸水的,轻软有弹性的
4. eternal [i(ː)ˈtəːnl] *adj.* 永远(久)的;永恒的，永存的
5. squeeze [skwiːz] *vt.* 榨，挤（出）
6. porou [ˈpɔːrəs] *adj.*【医】多孔的
7. odor [ˈəudə] [美]=odour *n.* 气味，名声
8. refined [riˈfaind] *adj.* 精炼(制)的，优雅的，讲究的，洗炼的;极微妙的，精密的
9. petroleum [piˈtrəuliəm] *n.* 石油
10. Euphrates [juːˈfreitiːz] *n.* 幼发拉底河[亚洲]
11. sediment [ˈsedimənt] *n.* 沉淀（物），沉积;【地质】沉积物
12. Carboniferous [ˌkɑːbəˈnifərəs] *adj.*【地质】石炭纪的，含碳的

Exercises

1. Put the following into Chinese.

(1) fossil fuel (2) be similar to (3) natural gas (4) a simple chemical compound

(5) crude oil (6) come from

2. Answer the following question, according to text.

(1) What is the coal made up of?

(2) How many kinds of major forms of fossil fuels are in the World?

3. Translate the paragraph into Chinese.

(1) Natural gas is lighter than air. Natural gas is mostly made up of a gas called methane. Methane is a simple chemical compound that is made up of carbon and hydrogen atoms. Its chemical formula is CH_4—one atom of carbon along with four atoms hydrogen. This gas is highly flammable.

(2) Oil is another fossil fuel. It was also formed more than 300 million years ago. Some scientists say that tiny diatoms are the source of oil. Diatoms are sea creatures the size of a pin head. They do one thing just like plants; they can convert sunlight directly into stored energy.

(3) Natural gas is lighter than air. Natural gas is mostly made up of a gas called methane. Methane is a simple chemical compound that is made up of carbon and hydrogen atoms. Its chemical formula is CH_4-one atom of carbon along with four atoms hydrogen. This gas is highly flammable.

Unit 8 The Mechanisms of Gaseous Fuels Combustion

Combustion is a chemical reaction during which a fuel is oxidized and a large quantity of energy is released. Air that enters a combustion chamber normally contains some water vapor (or moisture), which also deserves consideration. For most combustion processes, the moisture in the air and the H_2O that forms during combustion can also be treated as an inert gas, like nitrogen. At very high temperatures, however, some water vapor dissociates into H_2 and O_2 as well as into H, O, and OH. When the combustion gases are cooled below the dew-point temperature of the water vapor, some moisture condenses. It is important to be able to predict the dew-point temperature since the water droplets often combine with the sulfur dioxide that may be present in the combustion gases, forming sulfuric acid, which is highly corrosive.

During a combustion process, the components that exist before the reaction are called reactants and the components that exist after the reaction are products. Consider, for example, the combustion of 1 kmol of carbon with 1 kmol of pure oxygen, forming carbon dioxide,

$$C+O_2 \longrightarrow CO_2$$

Here C and O_2 are the reactants since they exist before combustion, and CO_2 is the product since it exists after combustion. Note that a reactant does not have to react chemically in the combustion chamber. For example, if carbon burned with air instead of pure oxygen, both sides of the combustion equation will include N_2. That is, the N_2 will appear both as a reactant and as a product.

Combustion of gaseous fuels occurs by the laws of branched chain reactions which were discovered by Soviet Academician N. N. Semenov and C. N. Hinshelwood. The conversion of the original substances to the final products passes through a sequence of reaction links which are connected in succession with one another and develop in the volume of a combustible mixture like the branches of a tree develop from its trunk. This results in the formation of the final reaction products and of even greater number of active centres which further ensure the development of the reaction in the confining volume.

Let us consider the mechanism of branched chain reactions, taking as an example the combustion of hydrogen in air. By the stoichiometric equation

$$2H_2+O_2 = 2H_2O$$

The rate of the reaction between molecules of the combustible substance

$$W_{H_2O}=k_0 e^{-\frac{E}{RT}} c_{H_2}^2 c_{H_2O} \qquad (1-11)$$

cannot be very large. Actually, however, combustion of hydrogen at temperatures above 500℃ is an explosive chain reaction proceeding at a very high rate. Indeed, according to N. N. Semenov, the beginning of the active reaction is preceded by the formation of active centres:

$$H_2+M^{\alpha} \longrightarrow 2H+M$$

$$H_2+O_2^{\alpha} \longrightarrow 2OH$$

Where M^{α} and O_2^{α} are active molecules which possess high energy levels in the volume.

Atoms and radicals formed by this mechanism actively enter the reactions with the surrounding molecules, i. e. chains of successive reactions develop which result in the formation of the final reaction products and ever greater number of active centres.

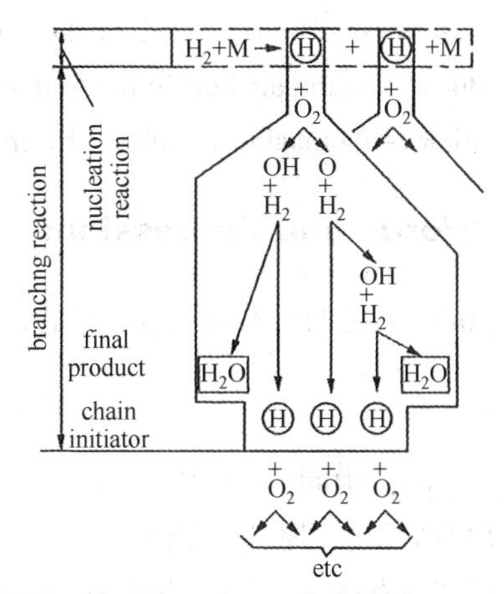

Fig. 1.16 Chain reaction cycle of hydrogen burning

○—chain reaction exciter; □—final product

Fig. 1.16 schematically shows the first cycle of this reaction. As may be seen, each of the active hydrogen atoms H that has given rise to a chain reaction has produced three new active centres, owing to which the reaction progressively develops in the volume confining the gas mixture. As the reaction products accumulate and the concentrations of the starting substances become lower, chains are disrupted more often in the volume and at the walls of the reactor:

$$H+H \longrightarrow H_2$$

$$OH+H \longrightarrow H_2O$$

The actual reaction rate is described by the equation:

$$W = 10^{-11}\sqrt{T}\exp\left(-\frac{7.54\times10^4}{RT}\right)c_{H}c_{O_2} \tag{1-12}$$

The decisive factors for the reaction rate are the concentrations of hydrogen atoms (reaction centres) and oxygen molecules, with the activation energy E' of the reaction between them being substantially lower than E in equation (1-11). Similar laws of chain reactions govern the combustion of carbon monoxide CO, methane CH_4 and other combustible gases.

It follows from the foregoing that a short time, the induction period, precedes the beginning of an active reaction, during which a sufficiently large quantity of active centres (atoms and radicals) accumulates in its reaction volume. During this period, the reaction is almost unnoticeable and thermal effect is negligible. After this period, the reaction rate increases due to the development of a large number of parallel reaction chains over the whole volume, until an equilibrium between the appearance and disappearance of active centres is established. The reaction then attains its maximum rate and will proceed at this rate, provided that fresh portions of starting substances are regularly supplied to the combustion zone.

Combustion of a gaseous fuel in a mixture with air occurs at a very high rate (a ready methane-air mixture burns in a volume of 10 m^3 in 0.1s). For this reason, the intensity of combustion of natural gas in steam boiler furnaces is limited by the speed at which it mixes with air in the burner, i. e.

by physical factors. The difficulties which arise when high flows of gas and air should be mixed thoroughly in a very short time in a burner are linked with the fact that the volume flow rates of the gas and air differ substantially, as approximately 10 m^3 of air are needed for the combustion of 1 m^3 of gas. For thorough intermixing, gas must be introduced into the air flow in the form of numerous fine jets and at a high rate. For the same purpose, the air flow is thoroughly turbulized by special swirling arrangements.

Words and Expressions

1. insufficient [ˌinsəˈfiʃənt] *adj.* 不足的，不够的；*n.* 不足
2. mechanism [ˈmekənizəm] *n.* 机理
3. methane [ˈmeθein] *n.* 甲烷
4. intermediate [ˌintəˈmi:djət] *adj.* 中间的；*n.* 中间物
5. valency [ˈveilənsi] *n.* （化合）价，（原子）价
6. medium [ˈmi:djəm] *n.* 媒体，方法，媒介；[ˈmi:diom] *n.* 媒体
7. induction [inˈdʌkʃən] *n.* 感应，诱导
8. sequence [ˈsi: kwəns] *n.* 连续，次序
9. confine [kənˈfain] *v.* 限制
10. concentration [ˌkɔnsenˈtreiʃən] *n.* 浓度，浓缩
11. thermal [ˈθəːməl] *adj.* 热的，热量的
12. parallel [ˈpærəlel] *adj.* 相似的，相同的；*n.* 相似处
13. approximately [əˈprɔksiˈmətli] *adv.* 大概，近于
14. swirl [swə:l] *n.* 旋涡，涡动

Exercises

1. Put the following into Chinese.

(1) branched chain reactions
(2) the rate of the reaction
(3) the concentrations of hydrogen atoms
(4) the activation energy
(5) in steam boiler furnaces

2. Answer the following question, according to text.

(1) During a combustion process, what is the reactants?
(2) What is called the induction period?

3. Translate the paragraph into Chinese.

(1) The minimum amount of air needed for the complete combustion of a fuel is called the stoichiometric or theoretical air.

(2) Combustion of a gaseous fuel in a mixture with air occurs at a very high rate (a ready methane-air mixture burns in a volume of 10 m^3 in 0.1s).

(3) It follows from the foregoing that a short time, the induction period, precedes the beginning of an active reaction, during which a sufficiently large quantity of active centres (atoms and radicals) accumulates in the its reaction volume.

Unit 9 The Combustion of Liquid Fuels and Solid Fuels

Although liquid hydrocarbon fuels are mixtures of many different hydrocarbons, they are usually considered to be a single hydrocarbon for convenience. For example, gasoline is treated as octane, C_8H_{18}, and the diesel fuel as dodecane, $C_{12}H_{26}$. Another common liquid hydrocarbon fuel is methyl alcohol, CH_3OH, which is also called methanol and is used in some gasoline blends. The gaseous hydrocarbon fuel natural gas, which is a mixture of methane and smaller amounts of other gases, is often treated as methane CH_4, for simplicity.

In the combustion of liquid fuels (petroleum, fuel oil), both the ignition and combustion temperatures (especially the latter) turn out to be higher than the boiling temperature of the individual fuel fractions. For this reason, liquid fuel first evaporates from the surface under the effect of the supplied heat, then its vapours are mixed with air, preheated to the ignition temperature and start burning. A stable flame forms at a certain distance from the surface of liquid fuel (0. 5 ~1 mm or more).

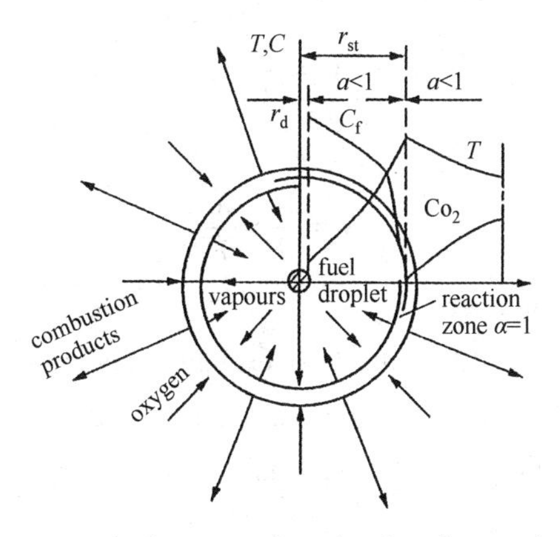

Fig. 1.17 Mechanism and combustion characteristics of a liquid fuel droplet

Fig. 1.17 schematically shows the combustion of a liquid fuel droplet in stagnant air. A vapour cloud forms around the droplet and diffuses into the environment, with the diffusion of oxygen of the air occurring in opposite direction. As a result, the stoichiometric relationship between the combustible gases and oxygen is established at a certain distance r_{st} from the droplet, i. e. the burning fuel vapours form a spherical combustion front around it. The magnitude of r_{st} is equal to 4 ~ 10 droplet radii, i. e. $r_{st}=4 \sim 10$rd, and depends heavily on the droplet size and the temperature in the combustion zone. In the zone where $r<r_{st}$, fuel vapours prevail, but their concentration decreases inversely with the distance from the liquid surface. The zone with $r>r_{st}$ contains primarily combustion products mixed with the oxygen that has diffused into the combustion zone. The highest temperature is established in the reaction zone. Although at both sides of this zone the temperature decreases gradually, its decrease is more intensive in the inside direction, i. e. on approaching the droplet, since some heat is spent there for heating fuel vapours.

Thus, the burning rate of a liquid fuel droplet is determined by the rate of evaporation from its surface, the rate of chemical reaction in the combustion zone, and the rate of oxygen diffusion to this zone. As stated earlier, the reaction rate in a gaseous medium is very high and cannot limit the total rate of combustion. The quantity of oxygen diffused through the spherical surface is proportional to the square of sphere diameter, and therefore, a slight removal of the combustion zone from the surface of the droplet (under oxygen deficiency) noticeably increases the mass flow rate of supplied oxygen. Thus, the rate of combustion of the droplet is mainly determined by evaporation from its surface. The combustion rate of liquid fuels is increased by atomizing the fuel just before burning, which substantially increases the total surface of evaporation. Besides all this, as the size of the droplets decreases, the intensity of evaporation per unit area of their surface increases. Fine liquid fuel droplets suspended in an air flow move at low Reynolds numbers, $Re \leqslant 4$. In such cases, the heat flow through a spherical surface is determined solely by the conductivity λ through the boundary layer, which is much thicker than the droplet diameter. Under such conditions, the heat-transfer efficient a is given by Sokolsky′s formula:

$$Nu = \frac{ad}{\lambda} = 2 \tag{1-13}$$

Whence

$$a = \frac{2\lambda}{d} = \frac{\lambda}{r} \tag{1-14}$$

where Nu is the Nusselt number.

As follows from formula (1-13), the heat exchange between a droplet and the surrounding medium increases as the size of the droplet decreases, i. e. with a decrease in its mass. It turns out that the evaporation time of a droplet is proportional to the square of its initial diameter.

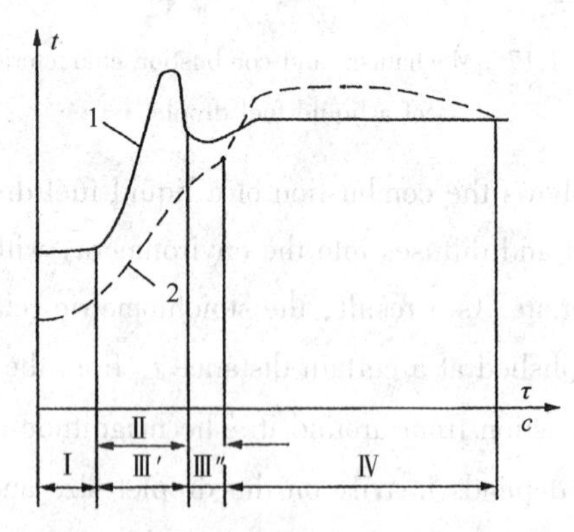

Fig. 1.18 Temperature conditions of burning of an individual solid fuel particle

1—temperature of gaseous medium around the particle; 2—particle temperature;

Ⅰ—thermal preparation zone; Ⅱ—zone of burning of volatiles; Ⅲ′—heating of coke particle due to burning of its volatiles;

Ⅲ″—heating of coke particle from an external source; Ⅳ—burning of coke particle

When combined with air in a furnace, pulverized coal first passes through the stage of thermal preparation (Fig. 1.18), which consists in the evaporation of residual moisture and separation of volatiles. Fuel particles are heated up to a temperature at which volatiles are evolved intensively (400~600℃) in a few tenths of a second. The volatiles are then ignited, so that the temperature around a coke particle increases rapidly and its heating is accelerated (Ⅲ′). The intensive burning of the volatiles (Ⅱ) takes up 0.2~0.5s. A high yield of volatiles (brown coal, younger coals, oil shales, peat), produces enough heat through combustion to ignite coke particles. When the yield of volatiles is low, the coke particles must be heated additionally from an external source (Ⅲ″). The final stage is the combustion of coke particles at a temperature above 800~1000℃ (Ⅳ). This is a heterogeneous process whose rate is determined by the oxygen supply to the reacting surface. The burning of a coke particle proper takes up the greater portion (1/2 to 2/3) of the total time of combustion which may constitute 1 to 2. 5 s, depending on the kind of fuel and the initial size of particles.

The reacting mechanism between carbon and oxygen seems to be as follows. Oxygen is adsorbed from the gas volume on the surface of particles and reacts chemically with carbon to form complex carbon-oxygen compounds of the type C_xO_y which then dissociate with the formation of CO_2 and CO. The resulting reaction at temperatures near 1200℃ can be written as follows

$$4C+3O_2 = 2CO+2CO_2$$

As has been established by experiment (L. Meyer, L. X. Khitrin), the ratio of the primary products. CO/CO_2, increases sharply with the increasing temperature of burning particles. For instance, the resulting equation at temperatures near 1700 ℃ can be written in the form:

$$3C+2O_2 = 2CO+CO_2$$

where the CO/CO_2 ratio is equal to two.

The primary reaction products are continuously removed from the surface of particles to the environment. In this process, carbon monoxide encounters the diffusing oxygen, which moves in the opposite direction, and reacts with it within the boundary gas film to be oxidized to CO_2, with the result that the concentration of supplied oxygen decreases sharply on approaching the surface of particles, while the concentration of CO_2 increases. At a high combustion temperature, carbon monoxide can consume all the oxygen supplied, which, consequently, will not reach the solid surface of particles. Under such conditions, the endothermic reduction reaction will occur on the surface of particles, i. e. CO_2 will be partially reduced to CO.

Thus, heterogeneous combustion of a carbon particle from its surface can be represented as a process embracing four subsequent reactions (according to A. S. Predvoditelev), two of which are the main ones:

$$C+O_2 = CO_2+q_1$$

$$2C+O_2 = 2CO+2q_2$$

the other two being secondary

$$2CO+O_2 = 2CO_2-2q_3$$

$$C+CO_2 = 2CO-q_4$$

where q is the thermal effect of a reaction, MJ/mol.

The thermal effect of the first reaction $q_1 = q_2 + q_3$, while $q_4 = 0.57\ q_3$. The latter equation implies that even when the endothermic reaction takes place, the temperature of combustion is maintained at a rather high level due to a higher heat evolution in the volume.

As follows from an analysis of these reactions, the combustion of carbon from the surface takes place with partial gasification (formation of CO and its afterburning in the volume). This process accelerates the burning-off of coke particles.

Words and Expressions

1. evaporate [iˈvæpəreit] *v.* 使蒸发, 使挥发
2. droplet [ˈdrɔplit] *n.* 小滴
3. spherical [ˈsferikəl] *adj.* 球的, 球形的
4. deficiency [diˈfiʃənsi] *n.* 缺乏, 不够
5. pulverize [ˈpʌlvəraiz] *v.* 将……粉碎
6. residual [riˈzidjuəl] *adj.* 残留, 剩余的
7. peat [piːt] *n.* 泥煤块
8. boundary [ˈbaundri] *n.* 界限, 边界
9. embrace [imˈbreis] *v.* 包含
10. endothermic [ˌendəʊˈθɜːmik] *v.* 吸热
11. concentration [ˌkɔnsenˈtreiʃən] *v.* 集中
12. established [isˈtæbliʃt] *adj.* 已制定的, 确定的
13. oil shales *n.* 石油页岩
14. diffuse [diˈfjuːz] *v.* 扩散, 散开
15. afterburner *n.* 喷射引擎等的加力燃烧室

Exercises

1. Put the following into Chinese.

(1) boundary layer (2) brown coal (3) Nusselt number
(4) younger coals (5) oil shales

2. Answer the following question, according to text.

(1) What is the Gasoline made up of?

(2) Heterogeneous combustion of a carbon particle from its surface can be represented as a process embracing four subsequent reactions, what are the conditions of the reactions?

3. Translate the paragraph into Chinese.

(1) Although liquid hydrocarbon fuels are mixtures of many different hydrocarbons, they are usually considered to be a single hydrocarbon for convenience. For example, gasoline is treated as octane, C_8H_{18}, and the diesel fule as dodecane, $C_{12}H_{26}$. Another common liquid hydrocarbon fuel is methyl alcohol, CH_3OH, which is also called methanol and is used in some gasoline blends. The gaseous hydrocarbon fuel natural gas, which is a mixture of methane and smaller amounts of other

gases, is often treated as methane CH_4, for simplicity.

(2) The reacting mechanism between carbon and oxygen seems to be as follows. Oxygen is adsorbed from the gas volume on the surface of particles and reacts chemically with carbon to form complex carbon-oxygen compounds of the type C_xO_y which then dissociate with the formation of CO_2 and CO. The reacting mechanism between carbon and oxygen seems to be as follows.

(3) The combustion rate of liquid fuels is increased by atomizing the fuel just before burning, which substantially increases the total surface of evaporation.

Unit 10 Theoretical and Actual Combustion Processes

It is often instructive to study the combustion of a fuel by assuming that the combustion is complete. A combustion process is complete if all the carbon in the fuel burns to CO_2, all the hydrogen burns to H_2O, and all the sulfur (if any) burns to SO_2. That is, all the combustible components of a fuel are burned to completion during a complete combustion process. Conversely, the combustion process is incomplete if the combustion products contain any unburned fuel or components such as C, H_2, CO, or OH.

Insufficient oxygen is an obvious reason for incomplete combustion, but it is not the only one. Incomplete combustion occurs even when more oxygen is present in the combustion chamber than is needed for complete combustion. This may be attributed to insufficient mixing in the combustion chamber during the limited time that the fuel and the oxygen are in contact. Another cause of incomplete combustion is dissociation, which becomes important at high temperatures.

Oxygen has a much greater tendency to combine with hydrogen than it does with carbon. Therefore, the hydrogen in the fuel normally burns to completion, forming H_2O, even when there is less oxygen than needed for complete combustion. Some of the carbon, however, ends up as CO or just as plain C particles (soot) in the products.

The minimum amount of air needed for the complete combustion of a fuel is called the stoichiometric or theoretical air. Thus, when a fuel is completely burned with theoretical air, no uncombined oxygen will be present in the product gases. The theoretical air is also referred to as the chemically correct amount of air, or 100 percent theoretical air. A combustion process with less than the theoretical air is bound to be incomplete. The ideal combustion process during which a fuel is burned completely with theoretical air is called the stoichiometric or theoretical combustion of that fuel. For example, the theoretical combustion of methane is

$$CH_4+2\,(O_2+3.76N_2)\longrightarrow CO_2+2H_2O+7.52N_2$$

Notice that the products of the theoretical combustion contain no unburned methane and no C, H_2, CO, OH, or free O_2.

In actual combustion processes, it is common practice to use more air than the stoichiometric amount to increase the chances of complete combustion or to control the temperature of the combustion chamber. The amount of air in excess of the stoichiometric amount is called excess air. The amount of excess air is usually expressed in terms of the stoichiometric air as percent excess air or percent theoretical air. For example, 50 percent excess air is equivalent to 150 percent theoretical air, and 200 percent excess air is equivalent to 300 percent theoretical air. Of course, the stoichiometric air can be expressed as 0 percent excess air or 100 percent theoretical air. Amounts of air less than the stoichiometric amount are called deficiency of air and are often expressed as percent deficiency of air. For example, 90 percent theoretical air is equivalent to 10 percent deficiency of air. The amount of air used in combustion processes is also expressed in terms of the equivalence ratio, which is the ratio of the actual fuel-air ratio to the stoichiometric fuel-air ratio.

Predicting the composition of the products is relatively easy when the combustion process is assumed to be complete and the exact amounts of the fuel and air used are known. All one needs to do in this case is simply apply the mass balance to each element that appears in the combustion equation, without needing to take any measurements. Things are not so simple, however, when one is dealing with actual combustion processes. For one thing, actual combustion processes are hardly ever complete, even in the presence of excess air. Therefore, it is impossible to predict the composition of the products on the basis of the mass balance alone. Then the only alternative we have is to measure the amount of each component in the products directly.

A commonly used device to analyze the composition of combustion gases is the Orsat gas analyzer. In this device, a sample of the combustion gases is collected and cooled to room temperature and pressure, at which point its volume is measured. The sample is then brought into contact with a chemical that absorbs the CO_2. The remaining gases are returned to the room temperature and pressure, and the new volume they occupy is measured. The ratio of the reduction in volume to the original volume is the volume fraction of the CO_2, which is equivalent to the mole fraction if ideal-gas behavior is assumed. The volume fractions of the other gases are determined by repeating this procedure. In orsat analysis the gas sample is collected over water and is maintained saturated at all times. Therefore, the vapor pressure of water remains constant during the entire test. For this reason the presence of water vapor in the test chamber is ignored and data are reported on a dry basis. However, the amount of H_2O formed during combustion is easily determined by balancing the combustion equation.

Words and Expressions

1. tendency [ˈtendənsi] *n.* 趋向，倾向
2. minimum [ˈminiməm] *adj.* 最小的，最低的；*n.* 最小值，最小化
3. behavior [biˈheivjə] *n.* 举止，行为
4. determine[diˈtəːmin] *v.* 决定，确定，测定，使下定决心，【律】使终止
5. stoichiometric *n.* 化学计量的，化学计算的
6. Orsat analysis 奥萨特（气体）分析（法）
7. bum [bʌm] *adj.* 无价值的
8. assume [əˈsjuːm] *vt.* 假定，设想，采取，呈现
9. entire [inˈtaiə] *adj.* 全部的，完整的，整个
10. assuming [əˈsjuːmiŋ] *adj.* 傲慢的，不逊的
11. equivalence [iˈkwivələns] *n.* 同等；【化】等价，等值

Exercises

1. Put the following into Chinese.

(1) combustion chamber (2) excess air (3) in the combustion equation

(4) the theoretical combustion (5) theoretical air

2. Answer the following question, according to text.

(1) What is the complete combustion process?

(2) What analyzer is a commonly used device to analyze the composition of combustion gases?

3. Translate the Paragraph into Chinese.

(1) A combustion process is complete if all the carbon in the fuel burns to CO_2, all the hydrogen bums to H_2O, and all the sulfur (if any) burns to SO_2. That is, all the combustible components of a fuel are burned to completion during a complete combustion process.

(2) Incomplete combustion occurs even when more oxygen is present in the combustion chamber than is needed for complete combustion.

(3) Oxygen has a much greater tendency to combine with hydrogen than it does with carbon. Therefore, the hydrogen in the fuel normally burns to completion, forming H_2O, even when there is less oxygen than needed for complete combustion. Some of the carbon, however, ends up as CO or just as plain C particles (soot) in the products.

(4) In this device, a sample of the combustion gases is collected and cooled to room temperature and pressure, at which point its volume is measured.

Unit 11 Pressure and Pressure Measurement

What is fluid pressure? Fluid pressure can be defined as the measure of force per-unit-area exerted by a fluid, acting perpendicularly to any surface it contacts (a fluid can be either a gas or a liquid, fluid and liquid are not synonymous). The standard SI unit for pressure measurement is the Pascal (Pa) which is equivalent to one Newton per square meter (N/m^2) or the kiloPascal (kPa) where 1 kPa = 1000 Pa. In the English system, pressure is usually expressed in pounds per square inch (psi). Pressure can be expressed in many different units including in terms of a height of a column of liquid.

Pressure measurements can be divided into three different categories: absolute pressure, gage pressure and differential pressure. Absolute pressure refers to the absolute value of the force per-unit-area exerted on a surface by a fluid. Therefore the absolute pressure is the difference between the pressure at a given point in a fluid and the absolute zero of pressure or a perfect vacuum. Gage pressure is the measurement of the difference between the absolute pressure and the local atmospheric pressure. Local atmospheric pressure can vary depending on ambient temperature, altitude and local weather conditions. The U. S. standard atmospheric pressure at sea level and 59℉ (20℃) is 14. 696 pounds per square inch absolute (psia) or 101. 325 kPa absolute (abs). When referring to pressure measurement, it is critical to specify what reference the pressure is related to. In the English system of units, measurement relating the pressure to a reference is accomplished by specifying pressure in terms of pounds per square inch absolute (psia) or pounds per square inch gage (psig). For other units of measure it is important to specify gage or absolute. The abbreviation 'abs' refers to an absolute measurement. A gage pressure by convention is always positive. A 'negative' gage pressure is defined as vacuum. Vacuum is the measurement of the amount by which the local atmospheric pressure exceeds the absolute pressure. A perfect vacuum is zero absolute pressure. Fig. 1. 19 shows the relationship between absolute, gage pressure and vacuum. Differential pressure is simply the measurement of one unknown pressure with reference to another unknown pressure. The pressure measured is the difference between the two unknown pressures. This type of pressure measurement is commonly used to measure the pressure drop in a fluid system. Since a differential pressure is a measure of one pressure referenced to another, it is not necessary to specify a pressure reference. For the English system of units this could simply be psi and for the SI system it could be kPa.

Most liquid and all gaseous materials in the process industries are contained within closed vessels. For the safety of plant personnel and protection of the vessel, pressure in the vessel is controlled. In addition, pressured is controlled because it influences key process operations like vapor-liquid equilibrium, chemical reaction rate, and fluid flow.

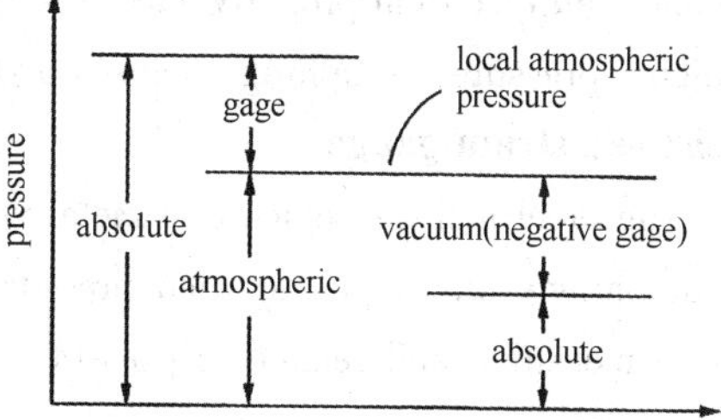

Fig. 1.19 Pressure term relationships

The following pressure sensors are based on mechanical principles, i. e., deformation based on force.

Bourdon

A bourdon tube is a curved, hollow tube with the process pressure applied to the fluid in the tube. The pressure in the tube causes the tube to deform or uncoil. The pressure can be determined from the mechanical displacement of the pointer connected to the Bourdon tube. Typical shapes for the tube are "C" (normally for local display), spiral and helical, shown as Fig. 1.20.

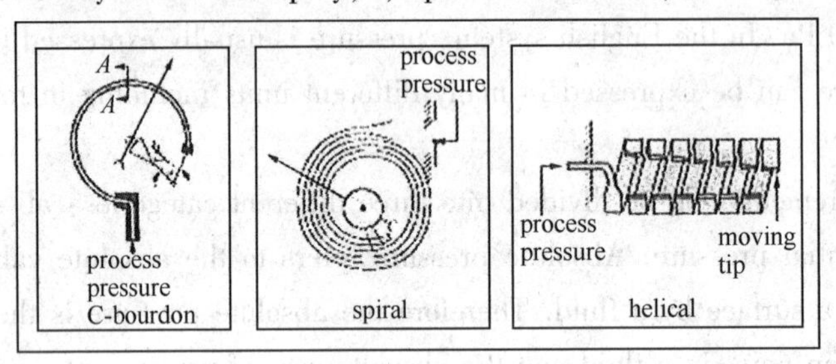

Fig. 1.20 Bourdon tube designs

Bellows

A bellows is a closed vessel with sides that can expand and contract, like an accordion. The position of the bellows without pressure can be determined by the bellows itself or a spring. The pressure is applied to the face of the bellows, and its deformation and its position depend upon the pressure.

Diaphragm

A diaphragm is typically constructed of two flexible disks, and when a pressure is applied to one face of the diaphragm, the position of the disk face changes due to deformation. The position can be related to pressure.

Diaphragms are popular because they require less space and because the motion (or force) they produced sufficient for operating electronic transducers. They also are available in a wide range of materials for corrosive service applications.

The following pressure sensors are based on electrical principles; some convert a deformation to a change in electrical property, others a force to an electrical property.

Capacitive or inductance

The movement associated with one of the mechanical sensors already described can be used to influence an electrical property such as capacitance affecting a measured signal. For example, under changing pressure a diaphragm causes a change in capacitance or inductance.

Resistive, strain gauge

The electrical resistance of a metal wire depends on the strain applied to the wire. Deflection of the diaphragm due to the applied pressure causes strain in the wire, and the electrical resistance can be measured and related to pressure.

Piezoelectric

A piezoelectric material, such as quartz, generates a voltage output when pressure is applied on it. Force can be applied by the diaphragm to a quartz crystal disk that is deflected by process pressure.

Words and Expressions

1. absolute pressure 绝对压力
2. gage pressure 表压，计示压力
3. differential pressure 压力落差
4. vacuum [ˈvækjuəm] *n.* 真空，空间
5. by convention 根据惯例
6. bourdon tube 布尔登管，波登管，弹簧管
7. bellows [ˈbeləuz] *n.* 波纹管，风箱
8. diaphragm [ˈdaiəfræm] *n.* 隔膜，膜片，光阑
9. capacitive [kəˈpæsitiv] *adj.* 电容的
10. inductance [inˈdʌktəns] *n.* 电感，感应系数
11. strain gauge 应变仪
12. piezoelectric [paiˌiːzəuiˈlektrik] *adj.* 【物】压电的

Exercises

1. Put the following into Chinese.

(1) absolute pressure (2) gage pressure (3) differential pressure
(4) pressure drop (5) strain gauge

2. Answer the following question, according to text.

(1) How many different categories pressure measurements can be divided into?
(2) How to determine the pressure in the bellows?
(3) In the text, how many pressure sensors are based on electrical principles?

3. Translate the paragraph 3,4,5,6 of the text into Chinese.

Unit 12 Introduction to Thermocouples

The thermocouple is one of the simplest of all sensors, as shown in Fig. 1.21. It consists of two wires of dissimilar metals joined near the measurement point. The output is a small voltage measured between the two wires.

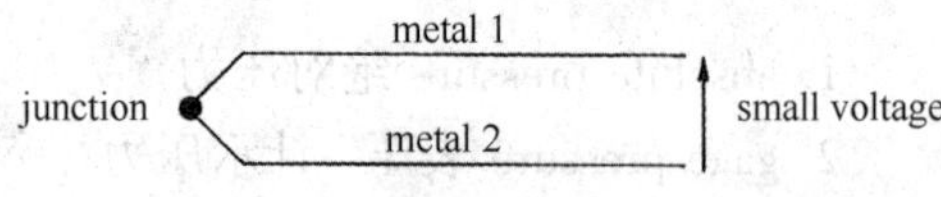

Fig. 1.21 The thermocouple

While appealingly simple in concept, the theory behind the thermocouple is subtle, the basics of which need to be understood for the most effective use of the sensor.

Thermocouple Theory

A thermocouple circuit has at least two junctions: the measurement junction and a reference junction. Typically, the reference junction is created where the two wires connect to the measuring device. This second junction it is really two junctions: one for each of the two wires, but because they are assumed to be at the same temperature (isothermal) they are considered as one (thermal) junction. It is the point where the metals change—from the thermocouple metals to what ever metals are used in the measuring device—typically copper.

The output voltage is related to the temperature difference between the measurement and the reference junctions. This is phenomena is known as the Seebeck effect. The Seebeck effect generates a small voltage along the length of a wire, and is greatest where the temperature gradient is greatest. If the circuit is of wire of identical material, then they will generate identical but opposite Seebeck voltages which will cancel. However, if the wire metals are different the Seebeck voltages will be different and will not cancel.

In practice the Seebeck voltage is made up of two components: the Peltier voltage generated at the junctions, plus the Thomson voltage generated in the wires by the temperature gradient, as shown in Fig. 1.22.

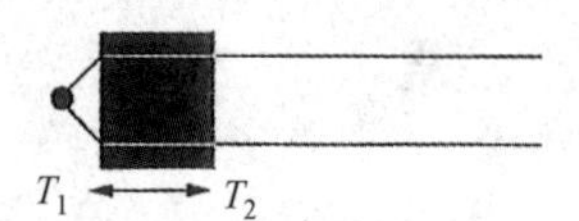

Fig. 1.22 Signal generated by temperature gradient

The Peltier voltage is proportional to the temperature of each junction while the Thomson voltage is proportional to the square of the temperature difference between the two junctions. It is the Thomson voltage that accounts for most of the observed voltage and non-linearity in thermocouple response.

Each thermocouple type has its characteristic Seebeck voltage curve. The curve is dependent on the metals, their purity, their homogeneity and their crystal structure. In the case of alloys, the ratio of constituents and their distribution in the wire is also important. These potential inhomogeneous characteristics of metal are why thick wire thermocouples can be more accurate in high temperature applications, when the thermocouple metals and their impurities become more mobile by diffusion.

The Practical Considerations of Thermocouples

The above theory of thermocouple operation has important practical implications that are well worth understanding:

(1) A third metal may be introduced into a thermocouple circuit and have no impact, provided that both ends are at the same temperature. This means that the thermocouple measurement junction may be soldered, brazed or welded without affecting the thermocouple′s calibration, as long as there is no net temperature gradient along the third metal.

Further, measured circuit metal (usually copper) is different to that of the thermocouple, then provided the temperature of the two connecting terminals is the same and known, the reading will not be affected by the presence of copper.

(2) The thermocouple′s output is generated by the temperature gradient along the wires and not at the junctions as is commonly believed. Therefore it is important that the quality of the wire be maintained where temperature gradients exists. Wire quality can be compromised by contamination from its operating environment and the insulating material. For temperatures below 400℃, contamination of insulated wires is generally not a problem. At temperatures above 1000℃, the choice of insulation and sheath materials, as well as the wire thickness, become critical to the calibration stability of the thermocouple.

The fact that a thermocouple′s output is not generated at the junction should redirect attention to other potential problem areas.

(3) The voltage generated by a thermocouple is a function of the temperature difference between the measurement and reference junctions. Fig. 1.23 shows the sketch of the modern thermocouple measurement. Traditionally the reference junction was held at 0℃ by an ice bath. The ice bath is now considered impractical and is replaced by a reference junction compensation arrangement. This can be accomplished by measuring the reference junction temperature with an alternate temperature sensor (typically an RTD or thermistor) and applying a correcting voltage to the measured thermocouple voltage before scaling to temperature.

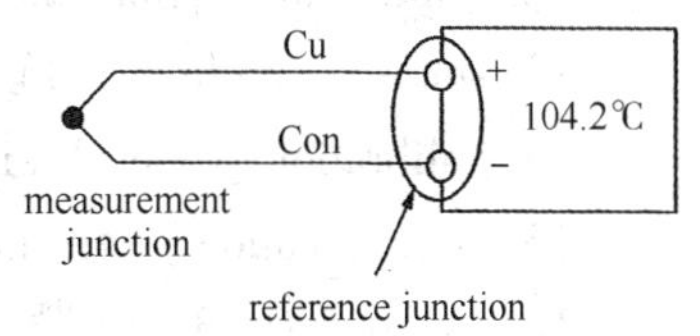

Fig. 1.23 Modern thermocouple measurement

The correction can be done electrically in hardware or mathematically in software. The software method is preferred as it is universal to all thermocouple types (provided the characteristics are known) and it allows for the correction of the small non-linearity over the reference temperature range.

The Advantages and Disadvantages of Thermocouples

Because of their physical characteristics, thermocouples are the preferred method of temperature measurement in many applications. They can be very rugged, immune to shock and vibration, are useful over a wide temperature range, are simple to manufactured, require no excitation power. There is no self heating and they can be made very small. No other temperature sensor provides this degree of versatility.

Thermocouples are wonderful sensors to experiment with because of their robustness, wide temperature range and unique properties.

On the down side, the thermocouple produces a relative low output signal that is non-linear.

These characteristics require a sensitive and stable measuring device that is able provide reference junction compensation and linearization. Also the low signal level demands that a higher level of care should be taken when installing to minimize potential noise sources.

Words and Expressions

1. dissimilar [diˈsimilə] *adj*. 不同的，相异的
2. appealingly [əˈpi：liŋ] *adv*. 有感染力，吸引人，有趣
3. subtle [ˈsʌtl] *adj*. 敏感的，微妙的，精细的
4. junction [ˈdʒʌŋkʃən] *n*. 连接，会合处，交叉点，接点
5. Seebeck effect 塞贝克效应
6. Peltier effect 珀耳帖效应
7. Thomson effect 汤姆逊效应
8. non-linearity [ˌnɔnliniˈæriti] *n*. 非线性
9. homogeneity [ˌhɔməudʒeˈni：iti] *n*. 同种，同质，同质性，均匀性
10. inhomogeneous [ˌinhəuməˈdʒi：niəs] *adj*. 不同类的，不同质的，非均匀的，不纯一的
11. mobile [ˈməubail] *adj*. 可移动的，易变的，机动的
12. implication [ˌimpliˈkeiʃən] *n*. 牵连，含义，暗示，推断
13. solder [ˈsɔldə] *vt*. 施以焊接，焊合，联结
14. calibration [ˌkæliˈbreiʃən] *n*. 口径测定，刻度，校准，测量
15. contamination [kənˌtæmiˈneiʃən] *n*. 玷污，污染，污染物
16. sheath [ʃi：θ] *n*. 鞘，护套，外壳
17. compensation [kɔmpenˈseiʃən] *n*. 补偿，赔偿

Exercises

1. Put the following into Chinese.

(1) Seebeck effect　(2) account for　(3) reference temperature
(4) ice bath　(5) on the down side

2. Answer the following question, according to text.

(1) What component Seebeck voltage is made up of?

(2) How we get the temperature of the reference junction temperature at present?

3. Translate the paragraph '***The practical considerations of thermocouples***' of the text into Chinese.

Unit 13 Flow Measurement

Flow measurement is critical to determine the amount of material purchased and sold, and in these applications, very accurate flow measurement is required. In addition, flows throughout the process should be regulated near their desired values with small variability; in these applications, good reproducibility is usually sufficient. Flowing systems require energy, typically provided by pumps and compressors, to produce a pressure difference as the driving force, and flow sensors should introduce a small flow resistance, increasing the process energy consumption as little as possible. Most flow sensors require straight sections of piping before and after the sensor. This requirement places restrictions on acceptable process designs, which can be partially compensated by straightening vanes placed in the piping. The sensors discussed in this subsection are for clean fluids flowing in a pipe; special considerations are required for concentrated slurries, flow in an open conduit, and other process situations.

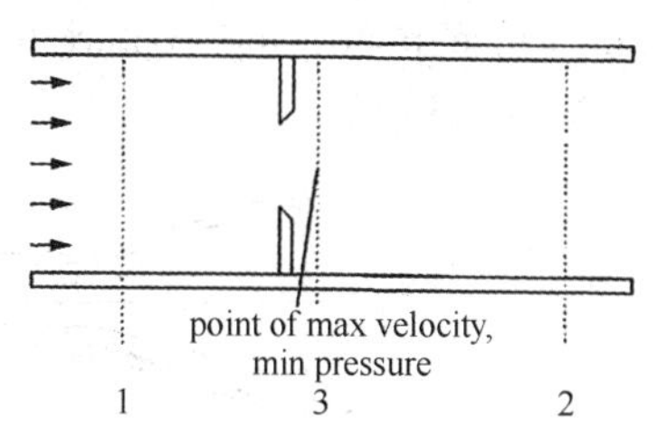

Fig. 1.24 Orifice flow meter

Several sensors rely on the pressure drop or head occurring as a fluid flows by a resistance. An example is given in Fig. 1.24. The relationship between flow rate and pressure difference is determined by the Bernoulli equation, assuming that changes in elevation, work and heat transfer are negligible.

Bernoulli's equation $$\frac{P_1}{\rho g} + \frac{1}{2g}v_1^2 = \frac{P_2}{\rho g} + \frac{1}{2g}v_3^2 + \sum f$$

Where $\sum f$ represents the total friction loss that is usually assumed negligible. This equation can be simplified and rearranged to give (Foust et. al, 1981; Janna, 1993) general head meter equation

$$F_1 = A_1 V_1 = C_{\text{meter}} Y A_3 \sqrt{\frac{2(P_1 - P_3)}{\rho(1 - A_3^2/A_1^2)}} \qquad (1)$$

The meter coefficient, C_{meter}, accounts for all non-idealities, including friction losses, and depends on the type of meter, the ratio of cross sectional areas and the Reynolds number. The compressibility factor, Y, accounts for the expansion of compressible gases. It is 1.0 for incompressible fluids. These two factors can be estimated from correlations (ASME, 1959; Janna, 1993) or can be determined through calibration. Equation (1) is used for designing head flow meters for specific plant operating conditions.

Some typical head meters are described briefly in the following.

Orifice: An orifice plate is a restriction with an opening smaller than the pipe diameter which is inserted in the pipe; the typical orifice plate has a concentric, sharp edged opening, as shown in Figure 1.24. Because of the smaller area the fluid velocity increases, causing a corresponding decrease in pressure. The flow rate can be calculated from the measured pressure drop across the orifice plate, P_1-P_3. The orifice plate is the most commonly used flow sensor, but it creates a rather large non-recoverable pressure due to the turbulence around the plate, leading to high energy consumption (Foust, 1981).

Venturi Tube: The venturi tube shown in Fig. 1.25 is similar to an orifice meter, but it is designed to nearly eliminate boundary layer separation, and thus form drag. The change in cross-sectional area in the venturi tube causes a pressure change between the convergent section and the throat, and the flow rate can be determined from this pressure drop. Although more expensive that an orifice plate; the venturi tube introduces substantially lower non-recoverable pressure drops (Foust, 1981).

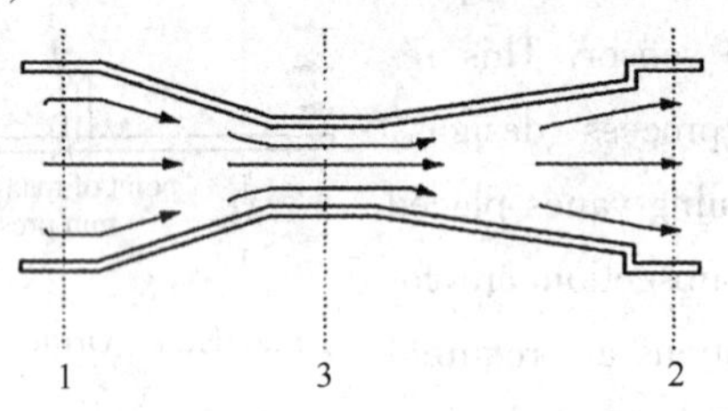

Fig. 1.25 Venturi flow meter

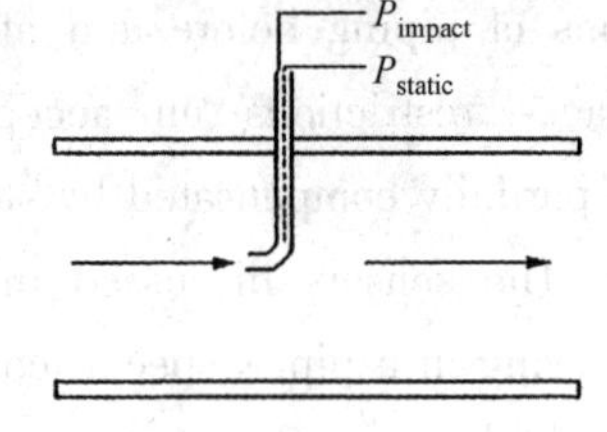

Fig. 1.26 Pitot flow meter

Flow Nozzle: A flow nozzle consists of a restriction with an elliptical contour approach section that terminates in a cylindrical throat section. Pressure drop between the locations one pipe diameter upstream and one-half pipe diameter downstream is measured. Flow nozzles provide an intermediate pressure drop between orifice plates and venturi tubes; also, they are applicable to some slurry systems.

Pitot tube and annubar: The pitot tube, shown in Fig. 1.26 below, measures the static and dynamic pressures of the fluid at one point in the pipe. The flow rate can be determined from the difference between the static and dynamic pressures which is the velocity head of the fluid flow. An annubar consists of several pitot tubes placed across a pipe to provide an approximation to the velocity profile, and the total flow can be determined based on the multiple measurements. Both the pitot tube and annubar contribute very small pressure drops, but they are not physically strong and should be used only with clean fluids.

The following flow sensors are based on physical principles other than head.

Turbine: As fluid flows through the turbine, it causes the turbine to rotate with an angular velocity that is proportional to the fluid flow rate. The frequency of rotation can be measured and used to determine flow. This sensor should not be used for slurries or systems experiencing large, rapid flow or pressure variation.

Vortex shedding: Fluid vortices are formed against the body introduced in the pipe. These vortices are produced from the downstream face in a oscillatory manner. The shedding is sensed using a thermistor and the frequency of shedding is proportional to volumetric flow rate.

Positive displacement: In these sensors, the fluid is separated into individual volumetric elements and the number of elements per unit time are measured. These sensors provide high accuracy over a large range. An example is a wet test meter.

Words and Expressions

1. reproducibility [riprə͵dju：sə'biliti] *n.* 重复能力，再现性
2. restriction [ris'trikʃən] *n.* 限制，限定，约束
3. vane [vein] *n.* (风车、螺旋桨等的) 翼，叶片
4. consumption [kən'sʌmpʃən] *n.* 消费，消费量，消耗量;耗量
5. conduit ['kɔndit] *n.* 导管，水管，沟渠
6. pressure drop 压降
7. pressure head 压头
8. correlation [͵kɔri'leiʃən] *n.* 相互关系，相关，关联
9. orifice flow meter 孔(板) 流速计
10. flow meter 流量计
11. orifice plate 孔板
12. substantially [səb'stænʃ(ə) li] *adv.* 实质上，本质上，大体上
13. convergent [kən'və：dʒənt] *adj.* 趋集于一点的，会聚性的，收敛的
14. Pitot tube 皮托管
15. velocity profile 速度剖面(图)，速度分布图，速度变化图
16. vortex shedding 涡旋脱落
17. shedding ['ʃediŋ] *n.* 脱落，蜕落，蜕落物，流出，散发，梭口，梭道
18. thermistor [θə：'mistə] *n.* 热敏电阻
19. wet test meter 湿式气体流量计,湿球气体流量计

Exercises

1. Put the following into Chinese.

(1) cross sectional areas　(2) Reynolds number　(3) Pitot tube

(4) orifice flow meter　(5) pressure head

2. Answer the following question, according to text.

(1) Which pressure drop is the smallest, orifice plate, Venturi Tube or Flow Nozzle?

(2) What is the most commonly used flow sensor?

3. Translate the paragraph ‘Orifice：,Venturi Tube：,Flow Nozzle：’ of the text into Chinese.

Unit 14 Introduction to Control Systems

Automatic control has played a vital role in the advance of engineering and science. In addition to its extreme importance in space-vehicle systems, missile-guidance systems, robotic systems, and the like, automatic control has become an important and integral part of modern manufacturing and industrial processes. For example, automatic control is essential in the numerical control of machine tools in the manufacturing industries, in the design of autopilot systems in the aerospace industries, and in the design of cars and trucks in the automobile industries. It is also essential in such industrial operations as controlling pressure, temperature, humidity, viscosity, and flow in the process industries.

Since advances in the theory and practice of automatic control provide the means for attaining optional performance of dynamic systems, improving productivity, relieving the drudgery of many routine repetitive manual operations, and more, most engineers and scientists must now have a good understanding of this field.

Historical Review

The first significant work in automatic control was James Watt's centrifugal governor for the speed control of a steam engine in the eighteenth century. Other significant works in the early stages of development of control theory were due to Minorsky, Hazen, and Nyquist, among many others. In 1922, Minorsky worked on automatic controllers for steering ships and showed how stability could be determined from the differential equations describing the system. In 1932, Nyquist developed a relatively simple procedure for determining the stability of closed-loop systems on the basis of open-loop response to steady-state sinusoidal inputs. In 1934, Hazen, who introduced the term servomechanisms for position control systems, discussed the design of relay servomechanisms capable of closely following a changing input.

During the decade of the 1940s, frequency-response methods (especially the Bode diagram methods due to Bode) made it possible for engineers to design linear closed-loop control systems that satisfied performance requirements. From the end of the 1940s to the early 1950s, the root-locus method due to Evans was fully developed.

The frequency-response and root-locus methods, which are the core of classical control theory, lead to systems that are stable and satisfy a set of more or less arbitrary performance requirements. Such systems are, in general, acceptable but not optimal in any meaningful sense. Since the late 1950s, the emphasis in control design problems has been shifted from the design of one of many systems that work to the design of one optimal system in some meaningful sense.

As modern plants with many inputs and outputs become more and more complex, the description of a modern control system requires a large number of equations. Classical control theory, which deals only with single-input-single-output systems, becomes powerless for multiple-input-multiple-output systems. Since about 1960, because the availability of digital computers made possible time-domain analysis of complex systems, modern control theory based on time-domain analysis and synthesis using state variables, has been developed to cope with the increased complexity of modern plants and the stringent requirements on accuracy, weight, and cost in military, space, and industrial applications.

During the years from 1960 to 1980, optimal control of both deterministic and stochastic systems, as well as adaptive and learning control of complex systems, were fully investigated. From 1980 to the present, developments in modern control theory centered around robust control, H_∞ control, and associated topics.

Now for digital computers have become cheaper and more compact, they are used as integral parts of control systems. Recent applications of modern control theory include such nonengineering systems as biological, biomedical, economic, and socioeconomic systems.

Definitions

Before we can discuss control systems, some basic terminologies must be defined.

Controlled Variable and Manipulated Variable. The controlled variable is the quantity or condition that is measured and controlled. The manipulated variable is the quantity or condition that is varied by the controller so as to affect the value of the controlled variable. Normally, the controlled variable is the output of the system. Control means measuring the value of the controlled variable of the system and applying the manipulated variable to the system to correct or limit deviation of the measured value from a desired value.

In studying control engineering, we need to define additional terms that are necessary to describe control systems.

Plants. A plant may be a piece of equipment, perhaps just a set of machine parts functioning together, the purpose of which is to perform a particular operation. In this book, we shall call any physical object to be controlled (such as a mechanical device, a heating furnace, a chemical reactor, or a spacecraft) a plant.

Processes. The Merriam-Webster Dictionary defines a process to be a natural, progressively continuing operation or development marked by a series of gradual changes that succeed one another in a relatively fixed way and lead toward a particular result or end; or an artificial or voluntary. progressively continuing operation that consists of a series of controlled actions or movements systematically directed toward a particular result or end. In this book we shall call any operation to be controlled a process. Examples are chemical, economic, and biological processes.

Systems. A system is a combination of components that act together and perform a certain objective. A system is not limited to physical ones. The concept of the system can be applied to abstract, dynamic phenomena such as those encountered in economics. The word system should, therefore, be interpreted to imply physical, biological, economic, and the like, systems.

Disturbances. A disturbance is a signal that tends to adversely affect the value of the output of a system. If a disturbance is generated within the system, it is called internal, while an external disturbance is generated outside the system and is an input.

Feedback Control. Feedback control refers to an operation that, in the presence of disturbances, tends to reduce the difference between the output of a system and some reference input and does so on the basis of this difference. Here only unpredictable disturbances are so specified, since predictable or known disturbances can always be compensated for within the system.

Words and Expressions

1. space vehicle *n.* 航天器
2. missile guidance 导弹的制导
3. autopilot [ˈɔːtəpailət] *n.* 自动驾驶仪
4. centrifugal governor 离心调速器(调节器)，离心式传感器
5. sinusoidal [ˌsainəˈsɔidəl] *adj.* 正弦曲线的
6. servomechanism[ˈsəːvəuˈmekənizəm] *n.* 自动驾驶装置，伺服机构(系统)，自动控制装置
7. relay servomechanism 继电器式伺服机构
8. root locus 根轨迹
9. biomedical [ˈbaiəuˈmedikəl] *adj.* 生物医学的
10. cope with *v.* 与……竞争，应付
11. stringent [ˈstrindʒənt] *adj.* 严厉的，迫切的
12. deterministic [diˌtəːmiˈnistik] *adj.* 确定性的
13. stochastic [stəuˈkæstik] *adj.* 随机的
14. controlled variable 控制量，可控变量，可调量，调节量
15. manipulated variable 被控变量
16. controller [kənˈtrəulə] *n.* 调节器;控制器，管理者
17. heating furnace 加热炉
18. spacecraft [ˈspeiskraːft] *n.* 航天飞船
19. progressively [prəˈgresivli] *adv.* 进步，先进，不断前进，累进，渐进，逐渐，渐次
20. feedback control 反馈控制

Exercises

1. Put the following into Chinese.

(1) cope with (2) controlled variable (3) manipulated variable

(4) heating furnace (5) feedback control

2. Answer the following question, according the text.

(1) When the root-locus method was developed?

(2) What are Controlled Variable and Manipulated Variable?

(3) What is the motive of feedback control?

3. Translate the paragraph 10, 11, 12, 13 of the text into Chinese.

Unit 15 Examples of Control Systems

Fig. 1.27 shows a schematic diagram of temperature control of an electric furnace. The temperature in the electric furnace is measured by a thermometer, which is an analog device. The analog temperature is converted to a digital temperature by an A/D converter. The digital temperature is fed to a controller through an interface. This digital temperature is compared with the programmed input temperature, and if there is any discrepancy (error), the controller sends out a signal to the heater, through an interface, amplifier, and relay, to bring the furnace temperature to a desired value.

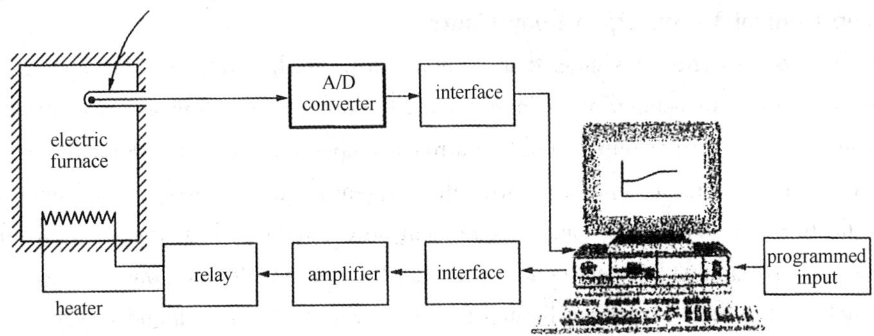

Fig. 1.27 Temperature control system

Consider the temperature control of the passenger compartment of a car. The desired temperature (converted to a voltage) is the input to the controller. The actual temperature of the passenger compartment must be converted to a voltage through a sensor and fed back to the controller for comparison with the input.

Fig. 1.28 is a functional block diagram of temperature control of the passenger compartment of a car. Note that the ambient temperature and radiation heat transfer from the sun, which are not constant while the car is driven, act as disturbances.

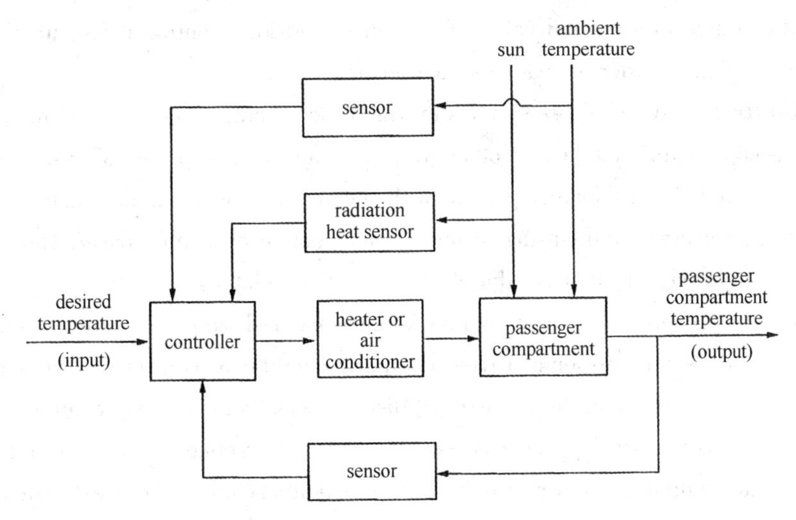

Fig. 1.28 Temperature control of passenger compartment of a car

The temperature of the passenger compartment differs considerably depending on the place where it is measured. Instead of using multiple sensors for temperature measurement and averaging the measured values, it is economical to install a small suction blower at the place where passengers normally sense the temperature. The temperature of the air from the suction blower is an indication of the passenger compartment temperature and is considered the output of the system.

The controller receives the input signal, output signal, and signals from sensors from disturbance sources. The controller sends out an optimal control signal to the air conditioner or heater to control the amount of cooling air or warm air so that the passenger compartment temperature is about the desired temperature.

Closed Loop Control Versus Open Loop Control

Feedback Control Systems. A system that maintains a prescribed relationship between the output and the reference input by comparing them and using the difference as a means of control is called a feedback control system. An example would be a room-temperature control system. By measuring the actual room temperature and comparing it with the reference temperature (desired temperature), the thermostat turns the heating or cooling equipment on or off in such a way as to ensure that the room temperature remains at a comfortable level regardless of outside conditions.

Feedback control systems are not limited to engineering but can be found in various non-engineering fields as well. The human body, for instance, is a highly advanced feedback control system. Both body temperature and blood pressure are kept constant by means of physiological feedback. In fact, feedback performs a vital function; it makes the human body relatively insensitive to external disturbances, thus enabling it to function properly in a changing environment.

Closed-loop Control Systems. Feedback control systems are often referred to as closed-loop control systems. In practice, the terms feedback control and closed-loop control are used interchangeably. In a closed-loop control system the actuating error signal, which is the difference between the input signal and the feedback signal (which may be the output signal itself or a function of the output signal and its derivatives and/or integrals), is fed to the controller so as to reduce the error and bring the output of the system to a desired value. The term closed-loop control always implies the use of feedback control action in order to reduce system error.

Open-Loop Control Systems. Those systems in which the output has no effect on the control action are called open-loop control systems. In other words, in an open-loop control system the output is neither measured nor fed back for comparison with the input. One practical example is a washing machine. Soaking, washing, and rinsing in the washer operate on a time basis. The machine does not measure the output signal, that is, the cleanliness of the clothes.

In any open-loop control system the output is not compared with the reference input. Thus, to each reference input there corresponds a fixed operating condition; as a result, the accuracy of the system depends on calibration. In the presence of disturbances, an open-loop control system will not perform the desired task. Open-loop control can be used, in practice, only if the relationship between the input and output is known and if there are neither internal nor external disturbances. Clearly, such systems are not feed-back control systems. Note that any control system that operates on a time basis is open loop. For instance, traffic control by means of signals operated on a time basis is another example of open-loop control.

Closed-Loop versus Open-Loop Control Systems. An advantage of the closed-loop control system is the fact that the use of feedback makes the system response relatively insensitive to external disturbances and internal variations in system parameters. It is thus possible to use relatively inaccurate and inexpensive components to obtain the accurate control of a given plant, whereas doing so is impossible in the open-loop case.

From the point of view of stability, the open-loop control system is easier to build because system stability is not a major problem. On the other hand, stability is a major problem in the closed-loop control system, which may tend to overcorrect errors and thereby can cause oscillations of constant or changing amplitude.

It should be emphasized that for systems in which the inputs are known ahead of time and in which there are no disturbances it is advisable to use open-loop control. Closed-loop control systems have advantages only when unpredictable disturbances and/or unpredictable variations in system components are present. Note that the Output power rating partially determines the cost, weight, and size of a control system. The number of components used in a closed-loop control system is more than that for a corresponding open-loop control system. Thus, the closed-loop control system is generally higher in cost and power. To decrease the required power of a system, open-loop control may be used where applicable. A proper combination of open-loop and closed-loop controls is usually less expensive and will give satisfactory overall system performance.

Words and Expressions

1. electric furnace 电炉
2. analog [ˈænəlɔg] *n*. 类似物，相似体
3. discrepancy [disˈkrepənsi] *n*. 相差，差异，矛盾，误差
4. compartment [kəmˈpaːtmənt] *n*. 间隔间，车厢
5. suction blower 吸风机，诱导通风机
6. prescribe [prisˈkraib]*v*. 指示，规定，处(方)，开(药)
7. feedback control 反馈控制
8. thermostat [ˈθəːməstæt] *n*. 自动调温器，温度调节装置
9. physiological [ˌfiziəˈlɔdʒikəl] *adj*. 生理学的，生理学上的
10. insensitive [inˈsensitiv] *adj*. 对……没有感觉的，感觉迟钝的
11. interchangeably *adv*. 可交换地，可替交地
12. washing machine 洗衣机
13. soak [səuk] *v*. 浸，泡，浸透
14. rinse [rins] *v*. (用清水) 刷，冲洗掉，漂净
15. washer [ˈwɔʃə] *n*. 洗衣人，垫圈，洗衣机
16. oscillation [ˌɔsiˈleiʃən] *n*. 摆动，振动

Exercises

1. Put the following into Chinese.

(1) electric furnace (2) desired value (3) suction blower

(4) feedback control (5) washing machine

2. Answer the following question, according to text.

(1) Which control system is generally higher in cost and power, closed-loop system or open-loop system?

(2) Whether feedback control systems and closed-loop control systems are interchangeable?

(3) How the temperature of an electric furnace is controlled?

3. Translate the paragraph 6,7,8,9 of the text into Chinese.

PART Ⅱ EQUIPMENT OF PYROLOGY

Unit 1 Boiler(1)

Introduction

Boilers use heat to convert water into steam for a variety of applications. Primary among these are electric power generation and industrial process heating. Steam has become a key resource because of its wide availability, advantageous properties and nontoxic nature. The steam flow rates and operating conditions can vary dramatically: from0. 1kg/s in one process to more than 1260kg/s in large electric power plants; from about 1bar and 100℃ in some heating applications to more than 310 bar and 593℃ in advanced cycle power plant.

Boiler classification

Modern steam generating systems can be classified by various criteria. These include end use, firing method, operating pressure, fuel and circulation method.

Utility steam generators are used primarily to generate electricity in large central power stations. They are designed to optimize overall thermodynamic efficiency at the highest possible availability. A key characteristics of newer units is the use of a reheater section to increase overall cycle efficiency.

Industrial steam generators generally supply steam to processes or manufacturing activities and are designed with particular attention to: ① process controlled (usually lower) pressure, ② high reliability with minimum maintenance, ③ use of one or more locally inexpensive fuels, especially process byproducts or wastes, and ④ low initial capital and minimum operating costs. On a capacity basis, the larger users of such industrial units are the pulp and paper industry, municipal solid waste reduction industry, food processing industry, petroleum/petrochemical industry, independent power producers and cogenerators, and some large manufacturing operations. Operating pressures range from 10 to 124 bar with saturated or superheated steam conditions.

Development of boiler

The modern 660MW coal-fired boiler has some 6000 tons of pressure parts which include 500 km of tubing, 3. 5 km of integral piping and 30000 tube butt welds. It is the culmination of some fifty years development and while the basic concept of pulverized fuel firing into a furnace lined with evaporator tubes, with the combustion gases then passing over convection superheater and heat recovery surface, has remained unchanged. The advancement of steam conditions, increasing in unit size and the properties of the fuel fired have required major changes in materials employed, fabrication techniques and operating procedures.

In the years immediately following the Second World War, it was customary to install in a power station, a greater number of boilers than turbines, the boilers feeding a range to which the turbines were connected. This arrangement reflected the inferior availability of boilers compared with turbines but increased in boiler availability in the late 1940s led to the acceptance of unitized boilers and turbines. The change to unitized boiler and turbine allowed reheat to become practicable and, with the availability of high temperature steels, there followed a continuous advance in steaming conditions to the current standard cycle of 2400 lbf/in^2, 165. 5bar, 568℃ with reheat to 568℃. To take full advantage of the more advanced steam conditions and to obtain the economies of size, the next fifteen years also saw a twenty-fold increase in unit size.

A utility normally procures plant from specialist manufacturers who have responsibility for design, manufacture, erection and commission. While the manufacturers carry out development of manufacturing processes and continuously update their design methods, and change in operating conditions and size necessarily results in a new plant being of a prototype nature. While some new features can be tested in advance of construction the only real test of a new boiler design is in operation and with its associated turbine and generator. The commercial success of a new design is proved over the whole projected life of the power station and utility, therefore, it has to balance the immediate economic advantages of a new design in terms of improved efficiency, reduce capital costs, etc, against the risks of poor availability, need for major modifications, etc, which might result from a new development. A utility normally purchases plant against generating needs and the repercussions of poor initial availability are not only being unable to meet load demand but also having to use costly plant to make up the shortfall. This period of major advance in steam cycle and unit size therefore required quite exceptional interaction with manufacturers in the design and fabrication area and development of operation and maintenance techniques to ensure that the economic gains did not prove illusory. These facets are now considered separately.

Other steam producing systems

A variety of additional systems also produce steam for power and process applications. These systems usually take advantage of low cost or free fuels, a combination of power cycles and processes, and recovery of waste heat in order to reduce overall costs. Examples of these include

Gas turbine combined cycle (*CC*) Advanced gas turbines with heat recovery steam generators as part of a bottoming cycle to use waste heat recovery and increase thermal efficiency.

Integrated gasification combined cycle (*IGCC*) Adding a coal gasifier to the CC to reduce fuel costs and to minimize airborne emissions.

Pressurized fluidized-bed combustion (*PFBC*) Including higher pressure combustion with gas cleaning and expansion of the combustion products through a gas turbine.

Blast furnace hood heat recovery Generating steam using the waste heat from a blast furnace.

Solar steam generator Using concentrators to collect and concentrate solar radiation and generate steam.

Words and Expressions

1. boiler [ˈbɔilə] *n.* 锅炉
2. reheater [ˈriːˈhiːtə] *n.* 再热器，回热器

3. petroleum [pi'trəuliəm] *n.* 石油
4. petrochemical [ˌpetrəu'kemikəl] *adj.* 石化的;*n.* 石化产品
5. cogenerator *n.* 热电联产机组
6. butt weld 对接焊
7. pulverized *adj.* 粉状的，磨成粉的
8. superheater ['sju:pəhi:tə] *n.* 过热器
9. turbine ['tə:bin] *n.* 涡轮机
10. lbf/in^2磅力每平方英寸，1 lbf/in^2 = 6894.76 Pa

Exercises

1. Put the following into Chinese.

(1) pulp and paper industry　(2) heat recovery　(3) fabrication techniques
(4) solar radiation　(5) integrated gasification combined cycle
(6) pressurized fluidized-bed combustion

2. Answer the following question, according to text.

(1) What is the function of boiler?
(2) Narrate the boiler classification standard.
(3) What is the advantage of the CC?

3. Translate the paragraph 1, 4, 5 into Chinese.

Unit 2 Boiler(2)

System arrangement and key components

Modern steam generators are a complex configuration of thermal-hydraulic (steam and water) sections which preheat and evaporate water, and superheat steam. These surfaces are arranged so that: ① the fuel can be burned completely and efficiently while minimizing emissions, ② the steam is generated at the required flow rate, pressure and temperature, and ③ the maximum amount of energy is recovered. A relatively simple coal-fired utility boiler is illustrated in Fig. 2.1. The major components in the steam generating and heat recovery system include

(1) Furnace and convection pass

(2) Steam superheaters (primary and secondary)

(3) Steam reheaters

(4) Boiler or steam generating bank

(5) Economizer

(6) Steam drum

(7) Attemperator and steam temperature control system

(8) Air heater

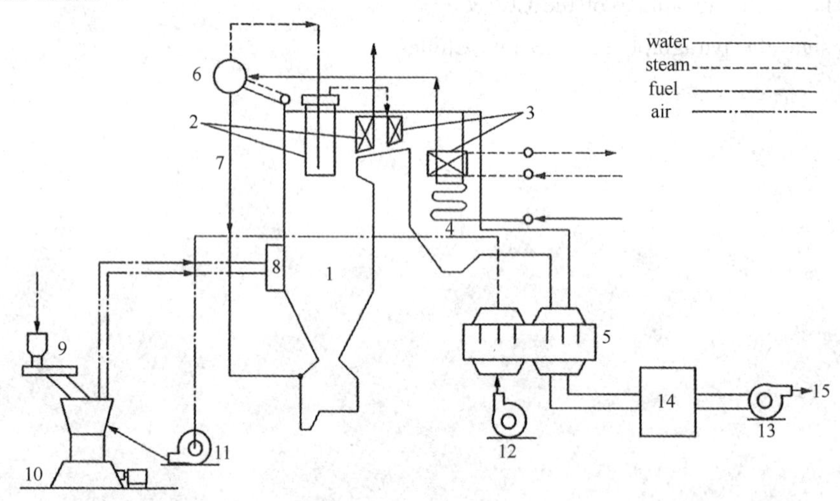

Fig. 2.1 Coal-fired Boiler

1—furnace; 2—superheater; 3—reheater; 4—economizer; 5—air heater; 6—drum; 7—down comer; 8—burner; 9—coal feeder; 10—pulverizer; 11—primary air fan; 12—forced draft fan; 13—induced draft fan; 14—precipitators; 15—exhaust

The furnace is a large enclosed space for fuel combustion and for cooling of the flue gas before it enters the convection pass. Excessive gas temperature leaving the furnace and entering the tube bundles could cause particle accumulation on the tubes or excessive tube metal temperature. The specific geometry and dimensions of the furnace are highly influenced by the fuel and type of combustion equipment. In this case, finely ground or pulverized coal is blown into the furnace where it burns in suspension. The products of combustion then rise through the upper furnace. The superheater, reheater and economizer surfaces are typically located in the horizontal and vertical downflow sections of the boiler.

In modern steam generators, the furnace and convection pass walls are composed of steam-cooled or water-cooled carbon steel or low alloy tube to maintain wall metal temperature within acceptable limits. These tubes are connected at the top and bottom by header, or manifolds. These headers distribute or collect the water, steam or steam-water mixture. The furnace wall tubes in most modern units also serve as key steam generating components or surfaces. The tubes are usually prefabricated into shippable membrane panels with openings for burners, observation doors, sootblowers (boiler cleaning equipment) and gas injection ports.

Superheaters and reheaters are specially designed inline tube bundles that increase the temperature of saturated steam. In general terms, they are simple-phase heat exchangers with steam flowing inside the tubes and the flue gas passing outside, general in crossflow. These critical components are manufactured from steel alloy material because of higher operating temperature. They are typically configured to help control steam outlet temperatures, keep metal temperatures below acceptable limits and control steam flow pressure loss.

The main difference between superheaters and reheaters is the steam pressure. In a typical drum boiler, the superheater outlet pressure might be 186 bar while the reheater outlet might be only 40 bar. The physical design and location of the surfaces depend upon the desired outlet temperatures, heat absorption, fuel ash characteristics and cleaning equipment. These surfaces can be either horizontal or vertical. The superheater and sometimes reheater are often divided into multiple sections to help control steam temperature and optimise heat recovery.

Economizers and air heaters perform a key function in providing high overall boiler thermal efficiency by recovering the low level, i. e. low temperature, energy from the flue gas before it is exhausted to the atmosphere. Economizers recover the energy by heating the boiler feedwater while air heaters heat the combustion air. Air heating also enhances the combustion of many fuels and ensuring ignition.

The economizer is a counterflow heat exchanger for recovering energy from the flue gas beyond the superheater and, if used, the reheater. It increases the temperature of the water entering the steam drum. The tube bundle is typically an arrangement of parallel horizontal serpentine tubes with the water flowing inside but in the opposition direction (counterflow) to the flue gas. Tube spacing is as tight as possible to promote heat transfer while permitting adequate tube surface cleaning and limiting flue gas side pressure loss. By design, steam is usually not generated inside these tubes.

The air heaters utilizes the heat in the boiler flue gases leaving the economizer to heat the combustion air and provide hot air for drying coal. An improvement of 1% in boiler efficiency is achieved for 22℃ rise in the coal combustion air temperature. The air outlet temperature limit in coal fired plant is dictated by the coal mill exit temperature and capacity of the tempering air system with the gas outlet temperature limited by considerations of fouling of the heat transfer surface and corrosion of downstream equipment.

On older boiler tubular or plate recuperators were generally used with which were large, difficult to clean and did not lend themselves to easy replacement of damaged heating surface. On all modern boilers regenerators are used, there are of two types, those supplied by James Howden&Co. Ltd. and those supplied by Davidson&Co. Ltd. The essential difference is that the Howden air heater is divided across the centre line with gas flow on one side and air on the other. The whole of the heating surface is contained within a rotating steel fabricated structure supported on a vertical shaft which rotates at approximately one rpm. The heat transfer surface is thus periodically heated by the flue gases and then gives up its heat to the incoming cold air. With the Davidson design the heat transfer surface is contained within a stationary structure. Above and below the heat transfer surface there are rotating hoods which direct the cold air through the heat transfer surface which has previously been heated by the flue gases. The cooled section is reheated as the hoods progress. The most significant feature of both types of regenerative air heaters, is the marked saving in space compared with recuperative designs.

Words and Expressions

1. economizer [iˈkɔnəmaizə] *n.* 省煤器
2. drum [ˈdrʌm] *n.* 汽包，锅筒
3. attemperator [əˈtempəˈreitə] *n.* 减温器，调温装置
4. air heater 空气预热器
5. down comer 下降管
6. precipitator 沉淀器，静电除尘器
7. suspension [səsˈpenʃən] *n.* 悬浮
8. header [hedə] *n.* 集箱，联箱
9. manifold [ˈmænifəuld] *n.* 集箱，集管，母管
10. sootblower *n.* 吹灰器
11. inline *n.* 顺列
12. serpentine tube 蛇形管，螺旋管
13. mill [mil] *n.* 磨煤机
14. recuperator [riˈkjuːpəreitə] *n.* 间壁式换热器
15. regenerator [riˈdʒenəreitə] *n.* 回热器，蓄热器

Exercises

1. Put the following into Chinese.

(1) tube bundles　　(2) saturated steam　　(3) counterflow heat exchanger

(4) pressure loss　　(5) flue gas　　(6) vertical shaft

2. Answer the following question, according text.

(1) What are the major for designing and locating in the boiler?

(2) What is the basis we design and locate the superheaters?

(3) What is the use of the air heaters?

3. Translate the paragraph 2, 4, 6 into Chinese.

Unit 3 Boiler(3)

Fuel Ash Corrosion

Fuel ash corrosion, also referred to as fireside corrosion or high temperature corrosion, is basically the oxidation of steels affected to a great or lesser extent by gaseous, molten or indeed solid components of the products of combustion of fossil fuels, i. e. coal and oil.

This type of corrosion is invariably associated with deposits formed on boiler tubes. These deposits consist of species resulting from the combustion of the fuel together with the corrosion products of the tube material. The composition of the deposits depends on the type and nature of the fuel and the conditions of combustion. In many instances chemical elements which are present in relatively small quantities, in some cases only in trace proportions, can concentrate on the tube surfaces and constitute the majority of the corrosive deposits.

During combustion the major constituents of fossil fuels, namely carbon and hydrogen, are converted to carbon dioxide and water-vapour. Therefore the flue gases mainly comprise carbon dioxide, water vapour, nitrogen(from the combustion air) and up to 4% oxygen (since excess air is used to ensure complete combustion) .

Both coal and oil contain organic and mineral impurities. Most of the impurity in coal is in the form of aluminum-silicates which leads to the formation of ash. Some of this ash, formed during combustion, falls to the bottom, i. e. hopper, of the furnace but most of it is carried by the combustion gases. Residual fuel oil contains very much smaller quantities of ash producing alumino-silicates, and all of this ash is carried by the combustion gases. It is the ash carried by combustion gases that forms the deposits on boiler tubes.

The main elements in coal which influence corrosion are sulphur, sodium, potassium and chlorine. Some of the sulphur in coal is organically bound as part of the coal matter, and some is present as iron pyrites thoroughly mixed with the coal. Sodium and potassium are present mainly in the alumino-silicate minerals.

Experience shows that chlorine content of the coal burnt in a boiler is a good indication of the corrosive potential of this coal. During the initial stages of the combustion of pulverized fuel particles the chlorine is released as hydrogen chloride gas, the hydrogen chloride concentration being directly proportional to the chlorine content of coal and the amount of excess air. Its role in the corrosion process is complex and possibly indirect. Recent work has shown that chlorine promotes the release of both sodium and potassium, and as there is always sufficient sulphur present, the released sodium and potassium compounds are converted to fusible sulphates. It is these fusible sulphates in the tube deposits that cause corrosion.

Furnace wall and superheater/reheater corrosion are influenced by fuel impurities in different ways, although in both cases the relatively stable and protective oxide scale that can be formed under ideal conditions by the oxidation of the metal surfaces by the oxygen and carbon dioxide present in the combustion gases is undermined or even prevented from forming in the first instance.

The protectiveness of the scale depends upon the rate of transport of the elemental species through it, its mechanical properties, and its adhesion to the metal surface.

Alloys with high chromium content produce thin and highly protective chromic scales. The adhesion of these scales to the metal substrate may be improved by additions of trace quantities of rare earth elements such as yttrium and cerium. If the strength of this bond is insufficient the differential thermal expansion of the scale and the metal can act as the driving force for scale separation. The propensity to void formation at or near the interface also influences the bond strength.

While there is some evidence to suggest that high chromium, i. e. greater than 20%. ferrite alloys may be more corrosion resistant than those alloys with an austenitic lattice structure. The austenitic alloys are preferred tube materials due to their high creep strength at elevated temperatures and better general mechanical integrity, e. g. they do not suffer from service embrittlement as do the high chromium ferrites.

Although the development of alloys with good oxidation resistance at high temperatures has been successful, this development has not eliminated the corrosion problems caused by sulphates and vanadates present in the ash deposits, which can gain direct access to the metal surface by undermining the otherwise protective scales, or departure from idea conditions preventing the formation of protective scales.

Furnace wall corrosion problems can be alleviated or criminated by the application of one or more of the following:

(1) Elimination of corrosive conditions by combustion system improvements or by burning coal with low, generally below 0.2% chlorine.

(2) Use of thicker tubes, faceted tubes, in the corrosion zone.

(3) Use of more corrosion resistant tubing, or the use of plasma spray coated tubing.

In the case of superheated and reheated corrosion the following remedies may be available or applied:

(1) Use of lower chlorine coal either naturally obtained or artificially washed and/or blended.

(2) By the control of temperature mal-distributions, prudent use of soot blowers and control of combustion to reduce differential slagging.

(3) Use of corrosion resistant metal shields over the affected areas of tubing.

(4) Replacement of tubing with more corrosion resistant tubing, such as co-extruded tubing, or the use of metal sprayed tubing(when this has been fully developed).

Words and Expressions

1. trace [treis] *n.* 微量,痕迹
2. aluminum-silicate *n.* 铝硅酸盐
3. hopper [ˈhɔpə] *n.* 斗,料斗,灰斗
4. residual fuel oil 残渣油
5. potassium [pəˈtæsiəm] *n.* 钠

6. chlorine [ˈklɔːriːn] *n.* 氯
7. pyrites [paiˈraitiːz] *n.*【矿】硫化铁矿
8. scale [skeil] *n.* 水垢,结垢
9. chromium [ˈkrəumjəm] *n.* 铬
10. yttrium [ˈitriəm] *n.* 钇
11. cerium [ˈsiriəm] *n.* 铈
12. austenitic *adj.* 奥氏体的
13. vanadate [ˈvænədeit] *n.* 钒酸盐(或酯)
14. plasma spray coated 等离子喷涂

Exercises

1. Put the following into Chinese.

(1) carbon dioxide (2) excess air (3) soot blower

(4) driving force (5) ferrite alloy

2. Answer the following question, according to text.

(1) The fossil fuels combustion produce flue gases, what are the main ingredients of the flue gases?

(2) Describe the main elements in coal which influence corrosion.

(3) How can reduce the furnace wall corrosion?

3. Translate the paragraph 1, 2, 13 into Chinese.

Unit 4 Steam Turbine(1)

Introduction

Turbine is a device that convert the energy deposited in a steam of fluid into mechanical energy.In the turbine, the fluid pass through a system of stationary passages and vanes which alternate with passages consisting of finlike blades attached to a rotor. Thus a torque is produced and exerted on the rotor blades. The torque makes the rotor turns, and work is extracted.Generally by the difference of work medium, turbines can be classified into four types: steam turbine, gas turbine, water turbine and wind turbine. On the whole the principles of these four types are same.

A steam turbine is a mechanical device that extracts thermal energy from pressurized steam, and converts it into useful mechanical work. Steam turbines are used in all of our major coal fired power stations to drive the generators or alternators, which produce electricity. The turbines themselves are driven by steam generated in 'Boilers' or 'Steam Generators' as they are sometimes called.

A steam turbine consists of a rotor resting on bearings and is enclosed in a cylindrical casing. Energy in the steam after it leaves the boiler is converted into rotational energy as it passes through the turbine. The turbine normally consists of several stages with each stage consisting of a stationary blade (or nozzle) and a rotating blade, which can be shown in Fig. 2.2. Stationary blades convert the potential energy of the steam (temperature and pressure) into kinetic energy (velocity) and direct the flow onto the rotating blades. The rotating blades convert the kinetic energy into forces, caused by pressure drop, which results in the rotation of the turbine shaft. The turbine shaft is connected to a generator, which produces the electrical energy. Thus a steam turbine could be viewed as a complex series of windmill-like arrangements, all assembled on the same shaft.

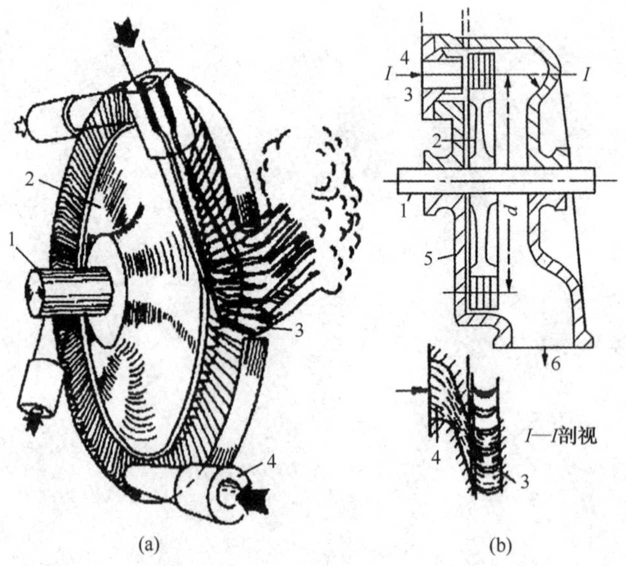

Fig. 2.2 The configuration of a single stage turbine

1—main shaft; 2— impeller; 3— rotating blade; 4— nozzle; 5— casing; 6—exhaust steam

For a simple steam turbine power plant, it consists of a heat source (boiler) that converts water to high-pressure steam. In the steam cycle, water is first pumped to medium or high pressure. It is then heated to the boiling temperature corresponding to the control, boiled (heated from liquid to vapor), and then most frequently superheated (heated to a temperature above that of boiling). Superheated steam enters the steam turbine through the governor valve. A turbine expands the pressurized steam to lower pressure and the steam is then exhausted either to a condenser at vacuum conditions or into an intermediate temperature steam distribution system that delivers the steam to the industrial or commercial application. The condensate water from the condenser or from the steam utilization system returns to the feedwater pump for continuation of the cycle. Fig. 2.3 shows the primary components of a boiler/steam turbine system.

Steam turbine has almost completely replaced the reciprocating piston steam engine, invented by Thomas Newcomen and greatly improved by James Watt, primarily because of its greater thermal efficiency and higher power-to-weight ratio. Also, because the turbine generates rotary motion, rather than requiring a linkage mechanism to convert reciprocating to rotary motion, it is particularly suited for driving an electrical generator — about 86% of all electric generation in the world is by use of steam turbines. Another use of steam turbines is in ships; their small size, low maintenance, light weight, and low vibration are compelling advantages. A steam turbine is efficient only when operating in the thousands of revolutions per minute (RPM) range while application of the power in propulsion applications may be only in the hundreds of RPM, which requires that expensive and precise reduction gears be used, although several ships, such as Turbinia, had direct drive from the steam turbine to the propeller shafts. This purchase cost is offset by much lower fuel and maintenance requirements and the small size of a turbine when compared to a reciprocating engine having an equivalent power. Most modern vessels now use either gas turbines or diesel engines, however, nuclear powered vessels such as aircraft carriers and nuclear submarines use steam turbines driving the propeller shaft through a reduction gearbox as the main part of their propulsion systems.

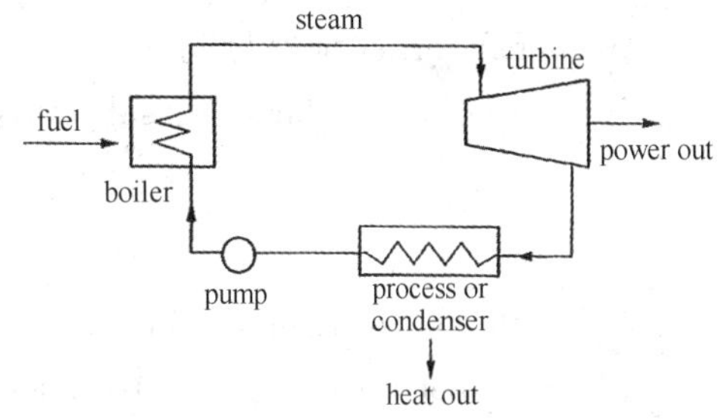

Fig. 2.3 Components of a boiler/steam turbine

Water turbines are the modern successors of simple waterwheels, which date back about 2,000 years. Today, the primary use of water turbines is for electric power generation. A water turbine uses the potential energy resulting from the difference in elevation between an upstream water reservoir and the turbine-exit water level (the tailrace) to convert this so-called head into work.

Wind turbines are the successors of windmills, which were important sources of power from the late Middle Ages through the 19th century. The energy available in wind can be extracted by a wind turbine to produce electric power or to pump water from wells.

In this chapter steam turbine and gas turbine will be briefly studied.

Words and Expressions

1. turbine [ˈtəːbin] *n.* 涡轮机
2. vane [vein] *n.* 静叶片
3. finlike [finlaik] *adj.* 鳍状的
4. torque [tɔːk] *n.* 扭矩，转矩
5. generator [ˈdʒenəreitə] *n.* 发电机，发生器
6. alternator [ˈɔːltə(ː) neitə] *n.* 交流发电机
7. condenser [kənˈdensə] *n.* 冷凝器，电容器
8. feedwater [ˈfiːdwɔːtə(r)] *n.* 给水
9. aircraft carrier 航空母舰
10. submarine [ˈsʌbmərːn, sʌbməˈriːn] *n.* 潜水艇，潜艇；*adj.* 水下的，海底的

Exercises

1. Put the following into Chinese.

(1) governor valve　　(2) condenser　　(3) working medium

(4) kinetic energy　　(5) diesel engines

2. Answer the following question, according to text.

(1) What is the working function of steam turbine?

(2) Narrate the working process of a simple steam turbine power plant according the text.

(3) What virtues does steam turbine have?

3. Translate the paragraph 3, 4, 5 into Chinese.

Unit 5 Steam Turbine(2)

Classification of steam turbines

Large steam turbines are complex machines that can be classified in various ways. They are given in Table 2.1.

Table 2.1 The classification of steam turbines

The basis of classification	Types of steam turbines
work function	impulse, reaction and combined impulse and reaction turbines
thermodynamic characteristic	condensing, back pressure, reheat, extraction turbines
flow direction	axial and radial flow turbine
applications	electric power generation, industrial and Marine turbines
steam conditions	low-pressure turbines, medium-pressure turbines, high-pressure turbines, supercritical pressure turbines
number of stages	single stage turbines and multi-stage turbines
steam entry configuration	full admission and partial admission turbines
number of flows	single flow, double flow and divided flow turbines

Among theses types, noncondensing or backpressure turbines are most widely used for process steam applications. The exhaust pressure is controlled by a regulating valve to suit the needs of the process steam pressure. These are commonly found at refineries, district heating units, pulp and paper plants, and desalination facilities where large amounts of low pressure process steam are available. The condensing turbine processes result in maximum power and electrical generation efficiency from the steam supply and boiler fuel. The power output of condensing turbines is sensitive to ambient conditions. Reheat turbines are also used almost exclusively in electrical power plants. In a reheat turbine, steam flow exits from a high pressure section of the turbine and is returned to the boiler where additional superheat is added. The steam then goes back into an intermediate pressure section of the turbine and continues its expansion. Extracting turbines are common in all applications. In an extracting turbine, steam is released from various stages of the turbine, and used for industrial process needs or sent to boiler feedwater heaters to improve overall cycle efficiency. Extraction flows may be controlled with a valve, or left uncontrolled.

The history and development of steam turbines

The first device that may be classified as a rudimentary steam turbine was the Aeolipile, created in the 1st century by Hero of Alexandria in Roman Egypt. It is little more than a toy, and no useful work was produced by it. In this Aeolipile, steam was supplied through a hollow rotating shaft to a hollow rotating sphere. It then emerged through two opposing curved tubes, just as water issues from a rotating lawn sprinkler.

The first practical steam turbine was invented much later in 1551 by Taqi al-Din in Ottoman Egypt, who described it as a prime mover for rotating a spit. Yet another steam turbine device was created by Italian Giovanni Branca in 1629. It was designed in such a way that a Jet of steam affected on blades extending from a wheel and caused it to rotate by the impulse principle. After 1784 when James Watt patented his steam engine, a number of reaction and impulse turbines were proposed, all adaptations of similar devices that operated with water. None were successful except for the units built by William Avery of the United States after 1837. About 50 percent turbines were built for sawmills, cotton gins, and Woodworking shops, and at least one was tried on a locomotive. But the turbines of that time had the disadvertage of low efficiencies matched those of contemporary steam engines, high noise levels, difficult speed regulation, and frequent needs for repairs, all of which led to their abandonment.

These early devices, however, were very different to the modern steam turbine, which was invented in 1884 by English engineer, Charles A. Parsons, whose first model was connected to a dynamo that generated 7. 5 kW of electricity. His patent was licensed and the turbine was scaled up shortly after by an American, George Westinghouse. The Parsons turbine turned out to be relatively easy to scale up. Within Parsons´ lifetime the generating capacity of a unit was scaled by a factor of about 10,000.

A number of other variations of turbines were developed that worked effectively with steam. From 1889 to 1897 de Laval built many turbines with capacities from about 15 to several hundred horsepower. The de Laval turbine accelerated the steam to full speed before running it against a turbine blade. The turbine was simpler, less expensive and, did not need to be pressure proof, and could operate with any pressure of steam.

After the first decade of the 20th century, steam turbines were the principal prime movers in central power stations. From 1900 to 1910, the capacity of steam turbine had increased from 1200 kilowatts to 30000 kilowatts. The out put of them far exceed that of the largest steam engine at that time.

The emergence of steam turbine impelled the development of electric power industry, by 20th the power of single turbine units had reached 10MW. From 1950, the economy development made the demand of electric power continuously increase, and the power of single turbine units as well as increased. The turbine with the power of 325~600MW had been developed in succession. By 1970, the power of the turbine had reached 1300MW. Now in many countries the power of single turbine units is 300~600MW.

Words and Expressions

1. aeolipile [iːt ˈɔlə, pail] *n.* 汽转球
2. Taqi al-Din 塔其汀
3. Ottoman[ˈɔtəmən] *adj* 土耳其帝国的;土耳其人的;土耳其民族的
4. Giovanni Branca 乔凡尼·白兰卡
5. William Avery 威廉·艾弗里
6. dynamo[ˈdainəməu] *n.* 发电机

7. scaled up 按比例增加【提高】
8. George Westinghouse 乔治. 威斯汀豪斯
9. de Laval turbine 单机冲动式汽轮机(拉瓦尔汽轮机)
10. variation [ˌvɛəri'eiʃən] *n*. 变更，变化，变异，变种，【音】变奏，变调

Exercises

1. Put the following into Chinese.

(1) backpressure turbines (2) condensing turbine (3) reheat turbines
(4) extracting turbines (5) heater (6) cycle efficiency

2. Answer the following question, according to text.

(1) How many kinds of steam turbine can be classified? What are they?
(2) Where can backpressure turbines be found?
(3) Narrate the history of steam turbine.

3. Translate the paragraph 2, 4 into Chinese.

Unit 6 Steam Turbine(3)

The degree of reaction of a turbo-machine stage is defined as the ratio of the static or pressure head change occurring in the rotor to the total change across the stage. On the basis of the degree of reaction the steam turbine can be classified into impulse and reaction steam turbine.

Impulse steam turbines

Impulse machine is the machine in which there is no change of static or pressure head of the fluid in the rotor that only cause energy transfer without any energy transformation. The energy transformation from pressure or static head to kinetic energy takes place only in fixed blades. In impulse turbine, the fixed nozzles orient the steam flow into high speed jets. These jets contain significant kinetic energy, which with the rotor blades, convert into shaft rotation as the steam jet changes direction. In impulse turbines, steam expansion only happens at nozzles. A pressure drop occurs in the nozzle. The pressure is the same when the steam enters the blade as it leaves the blade. As the steam flows through the nozzle, its pressure falls from steam chest pressure to condenser pressure (or atmosphere pressure) . Due to this relatively higher ratio of expansion of steam in the nozzle, the steam leaves the nozzle with a very high velocity. At a specific temperature and pressure steam has certain physical properties. The certain amount of heat or thermal energy contained within the steam with an increase of temperature or pressure the contained energy also increases. The flow of steam through a channel such as a nozzle reduces its thermal energy, however this decrease in thermal energy is equivalent the gain of kinetic energy. The thermal energy is converted from thermal to kinetic causing the steam to flow from high pressure, i. e. the steam chest, nozzle block, etc.. to an area of low pressure, i. e. the turbine casing. The steam leaving the moving blades still retains a large portion of the velocity it had after leaving the nozzle.

Since the rotor blade passages in an impulse turbine do not cause any acceleration of the fluid, the chances of its separation due to boundary layer growth on the blade surfaces are greater. On account of this, the rotor blade passages of the impulse machine suffer greater losses giving lower stage efficiencies. The loss of energy due to this higher exit velocity is commonly called the "carry over velocity" or "leaving loss."

Some examples of impulse machines are Banki turbine, Girard turbine, Pelton turbine and Turgo turbine.

Reaction steam turbines

Turbo machines or their stages in which changes in static or pressure head occur both in the rotor and stator blade passages are known as reaction machines or stages. Here the energy transformation occurs both in fixed as well as moving blades. The rotor experiences both energy transfer and transformation. In a reaction turbine the rotor blades themselves are arranged to form convergent nozzles. This type of turbine makes use of the reaction force produced as the steam accelerates through the nozzles formed by the rotor. Steam is directed onto the rotor by the fixed vanes of the stator. It leaves the stator as a jet that fills the entire circumference of the rotor. The steam then changes direction and increases its speed relative to the speed of the blades. A pressure drop occurs across both the stator and the rotor, with steam accelerating through the stator and decelerating through the rotor, with no net change in steam velocity across the stage but with a decrease in both pressure and temperature, reflecting the work performed in the driving of the rotor.

Half reaction machine has some special characteristics. Axial flow turbines and compressors with 50% reaction have symmetrical blades in their rotors and stators. It may be noted that the velocity triangles at the entry and exit of a 50% stage are also symmetrical. These types of turbines create large amounts of axial thrust, therefore, anti-friction thrust bearings are utilized.

Some examples of reaction machines are Hero's turbine, the lawn sprinkler and Parson's steam turbine.

Multi-stage steam turbine

In large steam turbines the difference of steam pressure between the boiler and the condenser's very large. If this was to be utilized in a single stage, a rotor of an impracticably large diameter at a very high speed would have to be used. This would create, besides manufacturing difficulties, serious strength and bearing problems.

In a typical larger power stations, the steam turbines are split into three separate parts, the first being the High Pressure (HP), the second the Intermediate Pressure (IP) and the third the Low Pressure (LP) part, where high, intermediate and low describe the pressure of the steam. After the steam has passed through the HP part, it is returned to the boiler to be re-heated to its original temperature although the pressure remains greatly reduced. The reheated steam then passes through the IP part and finally to the LP part of the turbine.

Multi-stage machines may employ only impulse or reaction stages or a combination of these. Impulse machines may utilize a large pressure drop in a number of pressure stages or a high kinetic energy in a number of velocity stages; a combination of pressure and velocity stages in impulse machines is also employed. In certain compressor applications it is profitable to use axial and radial stages in the same machine. Different stages may be mounted on one or more shafts.

Generally a multi-stage steam turbine consists of the following essential parts:

(1) A casing usually divided at the horizontal centerline, with the halves bolted together for ease of assembly and disassembly, and containing the stationary blade system.

(2) A rotor with the moving blades on wheels, and with bearing journals on the rotor.

(3) A bearing box in the casing, supporting the shaft.

(4) A governor and valve system for regulating the speed and power of the turbine by controlling the steam flow, and an oil system for lubrication of the bearings and a set of safety devices.

(5) A coupling of some soft to connect with the driven machine.

(6) Pipe connection to a supply of steam at the inlet, and to an exhaust system at the outlet of the casing.

Shaft Seals of steam turbine

The shaft seal on a turbine rotor consist of a series of ridges and groves around the rotor and its housing which present a long, tortuous path for any steam leaking through the seal. The seal therefore does not prevent the steam from leaking, merely reduces the leakage to a minimum. The leaking steam is collected and returned to a low-pressure part of the steam circuit.

Words and Expressions

1. reaction [ri(:)'ækʃən] *n.* 反应，反作用，反动(力)
2. impulse ['impʌls] *n.* 推动，刺激，冲动，推动力；*vt.* 推动
3. loss[lɔs] *n.* 损失，遗失，失败，输，浪费，错过；[军]伤亡，降低
4. Banki turbine 双击式水轮机
5. Girard turbine 吉拉德氏轮机
6. Pelton turbine 冲击式水轮机(立式)
7. Turgo turbine 斜击式水轮机
8. stator['steitə] *n.* 定子，固定片
9. symmetrical [si'metrikəl] *adj.* 对称的，均匀的
10. anti-friction 减少摩擦

Exercises

1. Put the following into Chinese.

(1) degree of reaction (2) Banki turbine (3) Girard turbine (4) Pelton turbine (5) Turgo turbine (6) multi-stage steam turbine (7) Shaft Seals

2. Answer the following question, according to text.

(1) How do you define impulse and reaction turbine?

(2) What is the reason for using multi-stage steam turbine?

(3) What is the use of Shaft Seals of steam turbine?

3. Translate the paragraph 2, 5, 10 into Chinese.

Unit 7 Steam Turbine(4)

Efficiency of steam turbine

Steam turbines have many stages, each of which generally consists of one row of stationary nozzles mounted on a casing and one row of moving curved blades mounted on a shaft. Each stage is designed to convert certain amount of thermal energy into mechanical work. Stage efficiency is defined as the ratio of mechanical work produced by the stage to the thermal energy available. Fig. 2.4 indicates stage process in an *h-s* diagram. In term of steam enthalpy, the stage efficiency is

$$\eta_s = (h_1 - h_2) / (h_1 - h_{2s}) \quad (2\text{-}1)$$

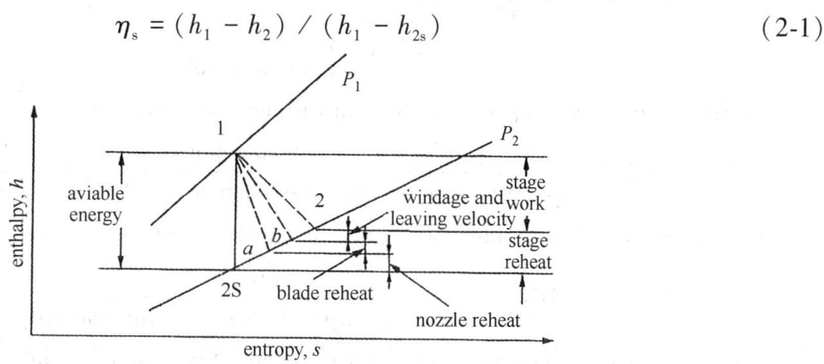

Fig. 2.4 An *h-s* diagram for turbine stage action

the value of η_s is always 0.84 ~ 0.90. Fig. 2. 4 also indicates friction effects on the enthalpy of steam. The frictional loss energy returns to steam in form of enthalpy. This phenomena is frequently called reheat.

To measure the performance of a steam turbine as a unit, the internal efficiency η_i, is frequently used. The term is defined as

$$\eta_i = \sum (\Delta W) / (\Delta H)_s \quad (2\text{-}2)$$

where $\sum (\Delta W)$ is the sum of internal work generated in all turbine stages; while $(\Delta H)_s$ is the isentropic enthalpy drop for the turbine. The range of η_i is 0.85 ~ 0.92.

In practice, the internal efficiency of a steam turbine does not include the loss at the turbine exhaust end, which occurs between the last stage of low-pressure turbine and the condenser inlet, and very much depends on the absolute steam velocity. The exhaust end loss generally includes actual leaving loss; gross hood loss; annulus-restriction loss; and turn-up loss. Under full-load condition the exhaust end loss is generally around 3% of the turbine′s available energy. One means of reducing the exhaust loss is to reduce the absolute steam velocity at the last stage by increasing the last stage blade length or the number of steam flows.

In addition to the internal efficiency and exhaust end loss, other parameters affecting the turbine performance are the gland and valve steam leakage losses (packing losses), mechanical turbine efficiency (bearing losses), and generator efficiency. In general, they constitute an additional small loss. In modern design mechanical efficiency is somewhere between 99% to 99. 5% and generator efficiency is 98. 5% to 99% for hydrogen-cooled generator.

Turbine Cycle Heat Balance

As discussed previously, the basic cycle used for electric power generation is the Rankine cycle, while the actural steam turbine cycle is much more complicated. With regenerative feedwater heating by turbine extraction steam and with interstage reheating. On the basis of energy and mass conservation principles, the general objective of heat and mass balances for the complicated turbine-cycle is to determine the system performance that is frequently measured in terms of turbine net heat rate or gross heat rate.

1) For the cycle arrangement with motor—driven boiler feedwater pump.

$$GHR = \frac{\text{heat input}}{\text{generator output}} \tag{2-3}$$

2) For the cycle arrangement with main turbine shaft-driven boiler feedwater pump.

$$GHR = \frac{\text{heat input}}{\text{generator output + turbines shaft output for feedwater pumps}} \tag{2-4}$$

3) For the cycle arrangement with auxiliary turbine shaft-driven boiler feedwater pumps.

$$GHR = \frac{\text{heat input}}{\text{generator output + auxiliary turbine output}} \tag{2-5}$$

The power plant design the turbine net heat rate(NHR) is frequently applied. The term is defined as

$$NHR = \frac{\text{heat input}}{\text{generator output-electric power for feedwater pumps}} \tag{2-6}$$

For the arrangement with motor-drive feedwater pumps, and

$$NHR = \frac{\text{heat input}}{\text{generator output}} \tag{2-7}$$

For the cycle arrangement with shaft-driven or auxiliary turbine-driven feedwater pumps. It must be pointed out that the heat input in these definitions means the amount of heat received by the steam in the steam generator.

A complete heat and mass balance generally needs the following parameters: a) Main steam pressure and temperature; b) Reheat steam temperature and pressure; c) Boiler and reheater pressure drops; d) Condenser pressure; e) Number of feedwater heater; f) Type of feedwater heaters; g) Heat drain disposals; h) Drain cooler approach; i) Heater terminal temperature difference; j) Steam extraction for auxiliary turbine; k) Steam extraction for industrial usage; l) Exhaust-end loss of low-pressure turbines.

Turbine-cycle heat and mass balance is time-consuming, and there are many possible cycle arrangements to be considered. Therefore, computer programs are frequently utilized for this particular purpose. However, the generalized heat balance program is expensive in its development and usually treated as proprietary material. One of the most famous computer programs developed for this purpose is the Advanced Generalized Heat Balance Program (AHBP), which consists of one main program and 23 subroutines. The main program has 11 sections: 1) Nomenclature; 2) Inputs and data file; 3) Printing and checking of important input data; 4) Calculation of turbine exhaust end conditions; 5) Calculation of steam extraction conditions; 6) Calculation of steam extraction rates; 7) Calculation of pump; 8) Calculation of turbine work and generator output; 9) Heat rate calculation; 10) Partial-load calculation; 11) Computer output printing.

Words and Expressions

1. enthalpy ['enθælpi, en'θælpi] *n.*【物】焓, 热函
2. isentropic [aisen'trɔpik] *adj.*【物】等熵的
3. loss [lɔs] *n.* 损失, 遗失, 失败, 输, 浪费, 错过,【军】伤亡, 降低
4. gross [grəus] *adj.* 总的, 毛重的;*n.* 总额
5. annulus ['ænjuləs] *n.* 环面
6. optimize ['ɔptimaiz] *vt.* 使最优化
7. reheat ['ri:'hi:t] *vt.* 再热
8. disposals [dis'pəuzəl] *n.* 处理, 处置, 布置, 安排, 配置, 支配
9. approach [ə'prəutʃ] *n.* 接近, 逼近, 走进, 方法, 步骤, 途径, 通路;*vt.* 接近, 动手处理;*vi.* 靠近
10. auxiliary [ɔ:g'ziljəri] *adj.* 辅助的, 补助的

Exercises

1. Put the following into Chinese.

(1) stage efficiency (2) internal efficiency (3) hood loss (4) interstage
(5) annulus-restriction (6) Rankine cycle (7) auxiliary turbine

2. Answer the following question, according to text.

(1) What is the definition of stage efficiency and internal efficiency?
(2) How can reduce the exhaust loss?
(3) List the parameter of a complete heat and mass balance.

3. Translate the paragraph 1, 3, 5, 8 into Chinese.

Unit 8 Steam Turbine(5)

Condenser

Condensers are employed in power plants to condense exhaust steam from turbines and in refrigeration plants to condense refrigerant vapors, such as ammonia and fluorinated hydrocarbons. The petroleum and chemical industries employ condensers for the condensation of hydrocarbons and other chemical vapors. In distilling operations, the device in which the vapor is transformed to a liquid state is called a condenser.

All condensers operate by removing heat from the gas or vapor; once sufficient heat is eliminated, liquefaction occurs. For some applications, all that is necessary is to pass the gas through a long tube(usually arranged in a coil or other compact shape) to permit heat to escape into the surrounding air. A heat-conductive metal, such as copper, is commonly used to transport the vapor. A condenser's efficiency is often enhanced by attaching fins(i. e., flat sheets of conductive metal) to the tubing to accelerate heat removal. Commonly, such condensers employ fans to force air through the fins and carry the heat away. In many cases, large condensers for industrial applications use water or some other liquid in place of air to achieve heat removal.

The steam turbine itself is a device to convert the heat in steam to mechanical power. The difference between the heat of steam per unit weight at the inlet to the turbine and the heat of steam per unit weight at the outlet to the turbine represents the heat which is converted to mechanical power. Therefore, the more the conversion of heat per pound or kilogram of steam to mechanical power in the turbine, the better is its efficiency. In order to operate an efficient close cycle, the condensing plant, cooling water (CW) system, and associated pumps must extract the maximum quantity of heat from the exhaust steam of the LP turbines. By condensing the exhaust steam of a turbine at a pressure below atmospheric pressure, the steam pressure drop between the inlet and exhaust of the turbine is increased, which increases the amount heat available for conversion to mechanical power. Thus we can conclude the primary functions of the condensing plant are: i) To provide the lowest economic heat rejection temperature for the steam cycle. ii) To convert the exhaust steam to water for reuse in the feed cycle. iii) To collect the useful residual heat from the drains of the turbine. iv) Heating plant, and other auxiliaries.

In addition to the condenser satisfying the primary functions, its design must also be capable of meeting the following objectives: Firstly to provide the turbine with the most economic back pressure consistent with the seasonal variations in CW temperature or the heat sink temperature of the CW system. Secondly to effectively prevent chemical contamination of the condensate either from CW leakage or from inadequate steam space gas removal and condensate deaeration.

Most of the heat liberated due to condensation of the exhaust steam is carried away by the cooling medium (water or air) used by the surface condenser. The Fig. 2.5 depicts a typical water-cooled surface condenser as used in power stations to condense the exhaust steam from a steam turbine driving an electrical generator as well in other applications. There are many fabrication design variations depending on the manufacturer, the size of the steam turbine, and other site-specific conditions.

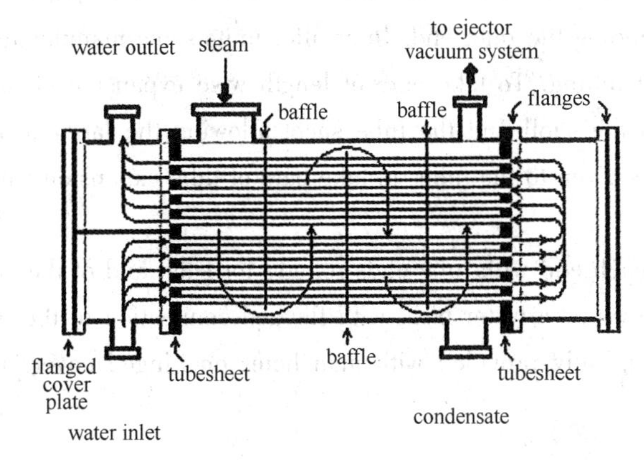

Fig. 2.5 Diagram of a typical water-cooled surface condenser

The shell is the condenser′s outermost body and contains the heat exchanger tubes. The shell is fabricated from carbon steel plates and is stiffened as needed to provide rigidity for the shell. When required by the selected design, intermediate plates are installed to serve as baffle plates that provide the desired flow path of the condensing steam. The plates also provide support that help prevent sagging of long tube lengths. At the bottom of the shell, where the condensate collects, an outlet is installed. In some designs, a sump (often referred to as the hot well) is provided. Condensate is pumped from the outlet or the hot well for reuse as boiler feedwater. For most water-cooled surface condensers, the shell is under vacuum during normal operating conditions.

For water-cooled surface condensers, the shell's internal vacuum is most commonly supplied by and maintained by an external steam jet ejector system(Fig. 2.6). Such an ejector system uses steam as the motive fluid to remove any non-condensible gases that may be present in the surface condenser. The venturi effect, which is a particular case of Bernoulli's principle, applies to the operation of steam jet ejectors. Motor driven mechanical vacuum pumps, such as liquid ring type vacuum pumps, are also popular for this service.

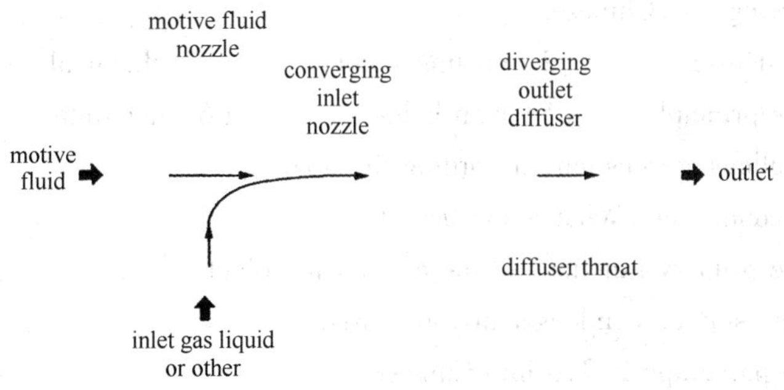

Fig. 2.6 Diagram of a typical modern injector or ejector

At each end of the shell, a tube sheet of sufficient thickness usually made of stainless steel is provided, with holes for the tubes to be inserted and rolled. The inlet end of each tube is also bellmouthed for streamlined entry of water. This is to avoid eddies at the inlet of each tube giving rise to erosion, and to reduce flow friction. Some makers also recommend plastic inserts at the entry of tubes to avoid eddies eroding the inlet end. In smaller units some manufacturers use ferrules to seal the tube ends instead of rolling. To take care of length wise expansion of tubes some designs have expansion joint between the shell and the tube sheet allowing the latter to move longitudinally. In smaller units some sag is given to the tubes to take care of tube expansion with both end water boxes fixed rigidly to the shell.

The tube sheet at each end with tube ends rolled, for each end of the condenser is closed by a fabricated box cover known as a water box, with flanged connection to the tube sheet or condenser shell. The water box is usually provided with man holes on hinged covers to allow inspection and cleaning.

Words and Expressions

1. refrigerant [riˈfridŋərənt] *adj.* 制冷的;*n.* 制冷剂
2. ammonia [ˈæməunjə] *n.*【化】氨, 氨水
3. fluorinated [ˈflu(:)ərineitd] *adj.* 氟化的
4. hydrocarbon [ˈhaidrəuˈka:bən] *n.* 烃, 碳氢化合物
5. distilling [disˈtiliŋ] *n.* 蒸馏(作用)
6. copper [ˈkɔpə] *n.* 铜
7. deaeration [diˌeiəˈreiʃən] *n.* 脱[排, 除]气(法), 脱泡(作用)
8. rigidity [riˈdʒiditi] *n.* 坚硬, 僵化, 刻板, 严格, 刚性, 硬度
9. sag [sæg] *v.* 松弛, 下陷, 下垂, (物价)下跌, 漂流;*n.* 下垂, 下陷, 物价下跌, 随风漂流, 垂度
10. bellmouthed [ˈbelmauðd] *adj.* 钟口形的, 漏斗口的, 承口的

Exercises

1. Put the following into Chinese.

(1) surface condensers (2) cooling water (3) chemical contamination

(4) Bernoulli's principle (5) man holes (6) hot well

2. Answer the following question, according the text.

(1) What is a condenser? What's the use of it?

(2) What is the primary functions of the condensing plant?

(3) Describe the surface condenser in you words.

3. Translate the paragraph 2,3,6 into Chinese.

Unit 9 Refrigeration

We all know from experience that heat flows in the direction of decreasing temperature, that is, from high-temperature regions to low-temperature ones. This heat-transfer process occurs in nature without requiring any devices. The reverse process, however, cannot occur by itself. The cycle that heat from a low-temperature region to a high-temperature one calls refrigeration.

Refrigeration was used by ancient civilizations when it was naturally available. Such natural sources of refrigeration were, of course, extremely limited in terms of location, temperatures, and scope. Means of producing refrigeration with machinery, called mechanical refrigeration, began to be developed in the 1850s. Today the refrigeration industry is a vast and essential part of any technological society, with yearly sales of equipment amounting to billions of dollars in the United States alone.

Uses of Refrigeration

It is convenient to classify the applications of refrigeration into the following categories: domestic, commercial, industrial, transport refrigeration and air conditioning. Sometimes transportation is listed as a separate category. Domestic refrigeration is used for food preparation and preservation, ice making, and cooling beverages in non-commercial areas such as offices and household throughout the world. Commercial refrigeration is used by retail outlets for preparing, holding and displaying frozen and fresh food and beverages for customer purchase. The transport refrigeration is aimed at transporting chilled or frozen goods. Industrial refrigeration in the food industry is needed in processing, preparation, and large-scale preservation. This includes use in food chilling and freezing plants, cold storage warehouses, breweries, and dairies, to name a few. Hundreds of other industries use refrigeration; among them are ice making plants, oil refineries, pharmaceuticals. Of course ice skating rinks need refrigeration. Refrigeration is also widely used in both comfort air conditioning for people and in industrial air conditioning. Industrial air conditioning is used to create the air temperatures, humidity, and cleanliness required for manufacturing processes. Computers require a controlled environment.

Methods of Refrigeration

Refrigeration, commonly spoken of as a cooling process is more correctly defined as the removal of heat from a substance to bring it to or keep it at a desirable low temperature, below the temperature of the surroundings. The most widespread method of producing mechanical refrigeration is called the vapor compression system. In this system a volatile liquid refrigerant is evaporated in an evaporator; this process results in a removal of heat (cooling) from the substance to be cooled. A compressor and condenser are required to maintain the evaporation process and to recover the refrigerant for reuse.

Another widely used method is called the absorption refrigeration system. In this process a refrigerant is evaporated (as with the vapor compression system), but the evaporation is maintained by absorbing the refrigerant in another fluid.

Other refrigeration methods are thermoelectric, steam jet, and air cycle refrigeration. These systems are used only in special applications and their functioning will not be explained here. Thermoelectric refrigeration is still quite expensive; some small tabletop domestic refrigerators are cooled by this method. Steam jet refrigeration is inefficient. Often used on ships in the past, it has been largely replaced by the vapor compression system. The air cycle is sometimes used in air conditioning of aircraft cabins. Refrigeration at extremely low temperatures, below about −200°F (−130℃), is called cryogenics. Special systems are used to achieve these conditions. One use of refrigeration at ultralow temperatures is to separate oxygen and nitrogen from air and to liquefy them.

Refrigeration Equipment

The main equipment components of the vapor compression refrigeration system are the familiar evaporator, compressor, and condenser. The equipment may be separate or of the unitary (also called serf-contained) type. Unitary equipment is assembled in the factory. The household refrigerator is a common example of unitary equipment. Obvious advantages of unitary equipment are that it is more compact and less expensive to manufacture if made in large quantifies.

There is a variety of commercial refrigeration equipment; each has a specific function. Reach-in cabinets, walk-in coolers, and display cases are widely used in the food service business. Automatic ice makers, drinking water coolers, and refrigerated vending machines are also commonly encountered equipment.

Air conditioning includes the heating, cooling, humidifying, dehumidifying, and cleaning (filtering) of air in internal environments. Occasionally it will be necessary to mention some aspects of air conditioning when we deal with the interface between the two subjects. A study of the fundamentals and equipment involved in air conditioning is nevertheless of great value even for those primarily interested in refrigeration.

Words and Expressions

1. refrigeration [riˌfridʒ'reiʃən] *n.* 制冷，冷藏，致冷，冷却
2. air conditioning 空气调节
3. comfort air conditioning 舒适性空调
4. industrial air conditioning 工艺性空调
5. evaporator [i'væpəreitə] *n.* 蒸发器
6. compressor [kəm'presə] *n.* 压缩机
7. condenser [kən'densə] *n.* 冷凝器
8. thermoelectric refrigeration 热电制冷
9. tabletop domestic refrigerator 家用台式电冰箱
10. steam jet refrigeration 蒸汽喷射式制冷
11. air cycle refrigeration 空气循环(膨胀)制冷
12. cryogenics [kraiəu'dʒeniks] *n.* 低温学
13. ultralow temperature 超低温

14. reach-in cabinet 大型陈列柜
15. dehumidify [ˌdiːhjuːˈmidifai] *vt.* 除湿，使干燥
16. display case 陈列柜，展示柜
17. walk-in cooler (可进入的)小型冷库

Exercises

1. Put the following into Chinese.

(1) mechanical refrigeration (2) domestic refrigeration (3) the vapor compression system (4) the absorption refrigeration system (5) refrigeration equipment

2. Answer the following question, according to text.

(1) What is refrigeration defined as?

(2) How many methods were mentioned in the text?

(3) What is the difference between the vapor compression system and the absorption refrigeration system?

3. Translate the paragraph 4,7,9 of the text into Chinese

Unit 10 Refrigerators and Heat Pumps

The transfer of heat from a low-temperature region to a high-temperature one requires special devices called refrigerators. Refrigerators are cyclic devices, and the working fluids used in the refrigeration cycles are called refrigerants. A refrigerator is shown schematically in Fig.2.7(a). Here Q_L is the magnitude of the heat removed from the refrigerated space at temperature T_L, Q_H is the magnitude of the heat rejected to the warm space at temperature T_H and $W_{net,in}$ is the net work input to the refrigerator. As discussed before, Q_L and Q_H represent magnitudes and thus are positive quantities.

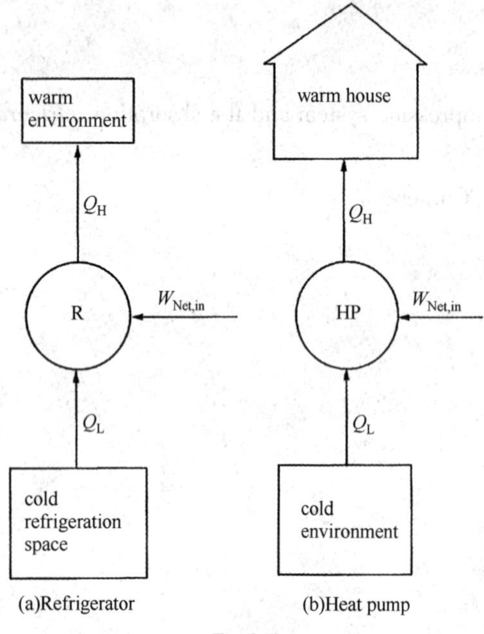

Fig.2.7

Another device that transfers heat from a low-temperature medium to a high-temperature one is the heat pump. Refrigerators and heat pumps are essentially the same devices; they differ in their objectives only. The objective of a refrigerator is to maintain the refrigerated space at a low temperature by removing heat from it. Discharging this heat to a higher-temperature medium is merely a necessary part of the operation, not the purpose. The objective of a heat pump, however, is to maintain a heated space at a high temperature. This is accomplished by absorbing heat from a low-temperature source, such as well water or cold outside air in winter, and supplying this heat to a warmer medium such as a house[Fig.2.7(b)].

The performance of refrigerators and heat pumps is expressed in terms of the coefficient of performance (COP), which was defined as

$$COP_R=\frac{\text{Desired output}}{\text{Required input}}=\frac{\text{Cooling effect}}{\text{Work input}}=\frac{Q_L}{W_{net,in}} \tag{2-8}$$

$$COP_{HP}=\frac{\text{Desired output}}{\text{Required input}}=\frac{\text{Heating effect}}{\text{Work input}}=\frac{Q_H}{W_{net,in}} \tag{2-9}$$

These relations can also be expressed in the rate form by replacing the quantities Q_L, Q_H and $W_{net,in}$ by $\dot{Q}_L$, $\dot{Q}_H$ and $\dot{W}_{net,in}$ respectively. Notice that both COP_R and COP_{HP} can be greater than 1. A comparison of Eqs. (2-8) and (2-9) reveals that

$$COP_{HP}=COP_R+1 \tag{2-10}$$

for fixed values of Q_L and Q_H. This relation implies that $COP_{HP}>1$ since COP_R is a positive quantity. That is, a heat pump will function, at worst, as a resistance heater, supplying as much energy to the house as it consumes. In reality, however, part of Q_H is lost to the outside air through piping and other devices, and COP_{HP} may drop below unity when the outside air temperature is too low. When this happens, the system normally switches to the fuel (natural gas, propane, oil, etc.) or resistance-heating mode.

The cooling capacity of a refrigeration system, that is, the rate of heat removal from the refrigerated space-is often expressed in terms of tons of refrigeration. The capacity of a refrigeration system that can freeze 1 ton(2000lbm) of liquid water at 0℃ (32℉) into ice at 0℃ in 24h is said to be 1 ton. One ton of refrigeration is equivalent to 211 kJ/min or 200 Btu/min. The cooling load of a typical $200m^2$ residence is in the 3ton (10kW) range.

Words and Expressions

1. refrigerator [riˈfridʒəreitə] *n.* 制冷机，冷冻机，冰箱，冷柜，冷气室
2. heat pump 热泵
3. refrigerant [riˈfridʒərənt] *n.* 制冷剂，冷冻剂，冷却介质
4. coefficient of performance 性能系数，经济性系数
5. COP_R 制冷系数，制冷机的性能系数
6. COP_{HP} 制热系数，热泵的性能系数
7. propane[ˈprəupein] *n.* 丙烷(可作制冷剂，制冷表示符号 R290)
8. lbm=pounds mass 磅(质量)
9. Btu=British thermal unit 英热量单位(=262 卡)

Exercises

1. Put the following into Chinese.

(1) net work (2) resistance-heating mode (3) cooling capacity of a refrigeration system

2. Answer the following question, according to text.

(1) What are called refrigerators?

(2) Why $COP_{HP}>1$?

3. Translate the paragraph 3, 4 of the text into Chinese.

Unit 11 The Ideal and the Actual Vapor-Compression Refrigeration Cycle

The Ideal Vapor-compression Refrigeration Cycle

The Schematic diagram for the basic vapor compression cycle and its *T-s* diagram are illustrated in Fig.2.8. Minimum components of this cycle include compressor, condenser, expansion valve and evaporator. The vapor-compression refrigeration cycle is the most widely used cycle for refrigerators, air-conditioning systems, and heat pumps. It consists of four processes:

1−2 Isentropic compression in a compressor

2−3 Constant-pressure heat rejection in a condenser

3−4 Throttling in an expansion device

4−1 Constant-pressure heat absorption in an evaporator

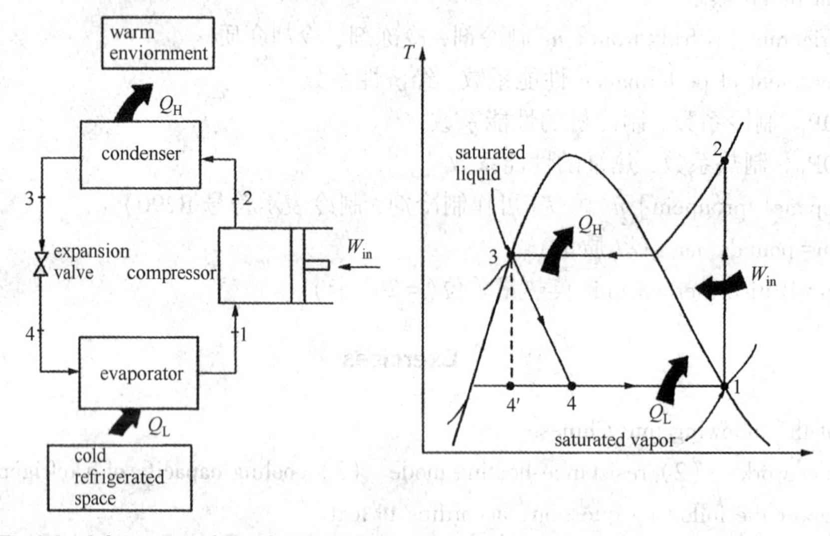

Fig.2.8 Schematic and *T-s* diagram for the ideal vapor-compression refrigeration cycle

In an ideal vapor-compression refrigeration cycle, the refrigerant leaves the evaporator as a low pressure, low temperature, saturated vapor and enters the compressor, where it is compressed reversibly and adiabatically (isentropic) to the condenser pressure. The temperature of the refrigerant increases during this isentropic compression process to well above the temperature of the surrounding medium. The refrigerant then enters the condenser as a high pressure, high temperature, superheated vapor at state 2 and leaves as saturated liquid at state 3 as a result of heat rejection to the surroundings. The temperature of the refrigerant at this state is still above the temperature of the surroundings. The saturated liquid refrigerant at state 3 is throttled to the evaporator pressure by passing it through an expansion valve or capillary tube. The temperature of the refrigerant drops below the temperature of the refrigerated space during this process. The refrigerant enters the evaporator at state 4 as a low quality saturated mixture, and it completely evaporates by absorbing heat from the refrigerated space. Then the low pressure, low temperature, saturated refrigerant vapor reenters the compressor, completing the cycle.

Remember that the area under the process curve on a $T-s$ diagram represents the heat transfer for internally reversible processes. The area under the process curve 4-1 represents the heat absorbed by the refrigerant in the evaporator, and the area under the process curve 2-3 represents the heat rejected in the condenser. A rule of thumb is that the COP improves by 2 to 4 percent for each ℃ the evaporating temperature is raised or the condensing temperature is lowered.

Another diagram frequently used in the analysis of vapor-compression refrigeration cycles is the $P-h$ diagram, as shown in Fig. 2. 9. On this diagram, three of the four processes appear as straight lines, and the heat transfer in the condenser and the evaporator is proportional to the lengths of the corresponding process curves.

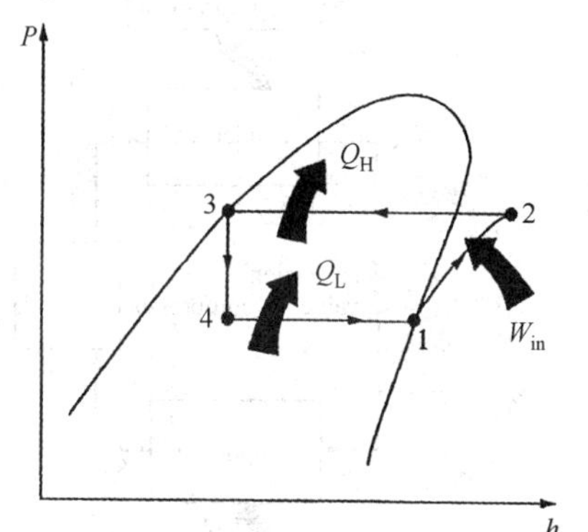

Fig.2.9 The P-h diagram of an ideal vapor compression refrigeration cycle

An energy balance and certain performance parameters can be derived from the first law of thermodynamics. Applying the steady flow equation for the first law to each of the components of the basic vapor compression cycle, the following relationships are derived:

$$W_{net,in}=h_2-h_1,\ q_L=h_1-h_4,\ q_H=h_2-h_3 \tag{2-11}$$

In applying the steady flow equation, the kinetic and potential energy changes of the refrigerant are usually small relative to the work and heat transfer terms, and therefore they can be neglected. Since the system is cyclic, the heat rejected in the condenser must be equal to the sum of the heat absorbed in the evaporator and the work of compression. Then the COP (Coefficient of Performance) of refrigerators and heat pumps operating on the vapor-compression refrigeration cycle can be expressed as:

$$COP_R=\frac{q_L}{W_{net,in}}=\frac{h_1-h_4}{h_2-h_1} \tag{2-12}$$

and

$$COP_{HP}=\frac{q_H}{W_{net,in}}=\frac{h_2-h_3}{h_2-h_1} \tag{2-13}$$

The Actual Vapor-compression Refrigeration Cycle

An actual vapor-compression refrigeration cycle differs from the ideal one in several ways, owing mostly to the irreversibilities that occur in various components. Two common sources of irreversibilities are fluid friction (causes pressure drops) and heat transfer to or from the surroundings. The $T-s$ diagram of an actual vapor-compression refrigeration cycle is shown in Fig.2.10.

In the ideal cycle, the refrigerant leaves the evaporator and enters the compressor as saturated vapor. Actually, it can not control the state of the refrigerant so precisely. Instead, however, it is easier to design the system so that the refrigerant is slightly superheated at the compressor inlet. This slight overdesign ensures that the refrigerant is completely vaporized when it enters the compressor. Also, the line connecting the evaporator to the compressor is usually very long; thus the pressure drop caused by fluid friction and heat transfer from the surroundings to the refrigerant can be very significant. The result of superheating, heat gain in the connecting line, and pressure drops in the evaporator and the connecting line is an increase in the specific volume, thus an increase in the power input requirements to the compressor since steady-flow work is proportional to the specific volume.

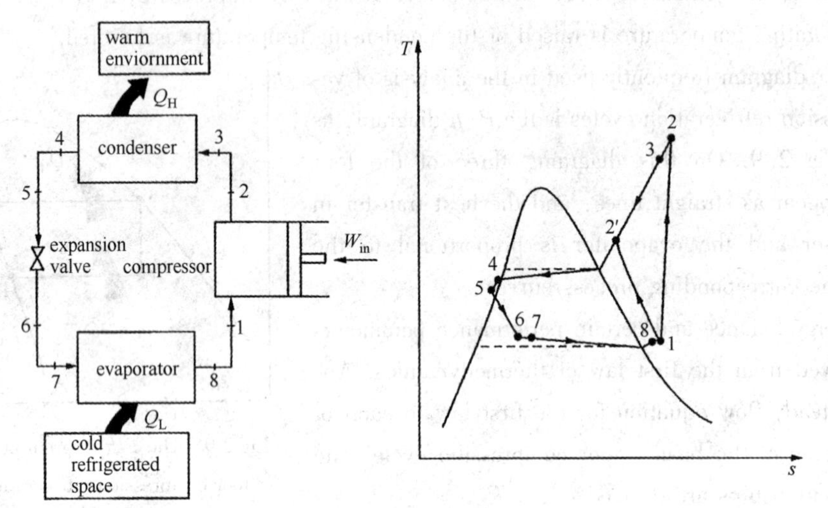

Fig.2.10 Schematic and T-s diagram for the actual vapor-compression refrigeration cycle

The compression process in the ideal cycle is internally reversible and adiabatic, and thus isentropic. The actual compression process, however, will involve frictional effects, which increase the entropy, and heat transfer, which may increase or decrease the entropy, depending on the direction. Therefore, the entropy of the refrigerant mayincrease or decrease during an actual compression process, depending on which effects dominate. The compression process may be even more desirable than the isentropic compression process since the specific volume of the refrigerant and thus the work input requirement are smaller in this case. Therefore, the refrigerant should be cooled during the compression process whenever it is practical and economical to do so.

In the ideal case, the refrigerant is assumed to leave the condenser as saturated liquid at the compressor exit pressure. In actual situations, however, it is unavoidable to have some pressure drop in the condenser as well as in the lines connecting the condenser to the compressor and to the throttling valve. Also, it is not easy to execute the condensation process with such precision that the refrigerant is a saturated liquid at the end, and it is undesirable to route the refrigerant to the throttling valve before the refrigerant is completely condensed. Therefore, the refrigerant is subcooled somewhat before it enters the throttling valve. And that the refrigerant in this case enters the evaporator with a lower enthalpy and thus can absorb more heat from the refrigerated space. The throttling valve and the evaporator are usually located very close to each other, so the pressure drop in the connecting line is small.

Words and Expressions

1. throttle device 节流装置，节流设备
2. expansion valve 膨胀阀
3. capillary tube 毛细管
4. isentropic [aisen'trɔpik] *adj.* 等熵的
5. saturated ['sætʃəreitid] *adj.* 饱和的，渗透的
6. superheated [ˌsju:pə'hi:tid] *adj.* 过热的
7. freezer compartment 低温格
8. kinetic [kai'netik] *adj.* (运)动的，动力(学)的
9. potential energy 势能，位能
10. unit-mass 单位质量
11. specific volume 比容，比体积
12. overdesign ['əuvədiˌzain] *vt.* & *n.* 超安全标准设计
13. adiabatic [ˌædiə'bætik] *adj.* 绝热的，隔热的
14. entropy ['entrəpi] *n.* 熵
15. subcool ['sʌbʌ'ku:l] *vt.* 使过冷，使低温冷却
16. enthalpy ['enθælpi, en'θælpi] *n.* 焓

Exercises

1. Put the following into Chinese.

(1) the ideal vapor-compression refrigeration cycle (2) constant-pressure heat rejection (3) a rule of thumb (4) steady-flow devices (5) the compressor exit pressure

2. Answer the following question, according to text.

(1) Please talk about the process of the ideal vapor-compression refrigeration cycle in English.

(2) What is the difference between the ideal vapor-compression refrigeration cycle and the actual vapor-compression refrigeration cycle?

3. Translate the paragraph 5,10 of the text into Chinese.

Unit 12 Innovative Vapor-Compression Refrigeration Systems

The simple vapor-compression refrigeration cycle discussed above is the most widely used refrigeration cycle, and it is adequate for most refrigeration applications. The ordinary vapor-compression systems are simple, inexpensive, reliable, and practically maintenance-free. However, for large industrial applications efficiency, not simplicity, is the major concern. Also, for some applications the simple vapor-compression refrigeration cycle is inadequate and needs to be modified. We shall now discuss a few such modifications and refinements.

Multistage Compression Refrigeration Systems

When the fluid used throughout the cascade refrigeration system is the same, the heat exchanger between the stages can be replaced by a mixing chamber (called a flash chamber) since it has better heat transfer characteristics. Such systems are called multistage compression refrigeration systems. A two-stage compression refrigeration system is shown inFig.2.11.

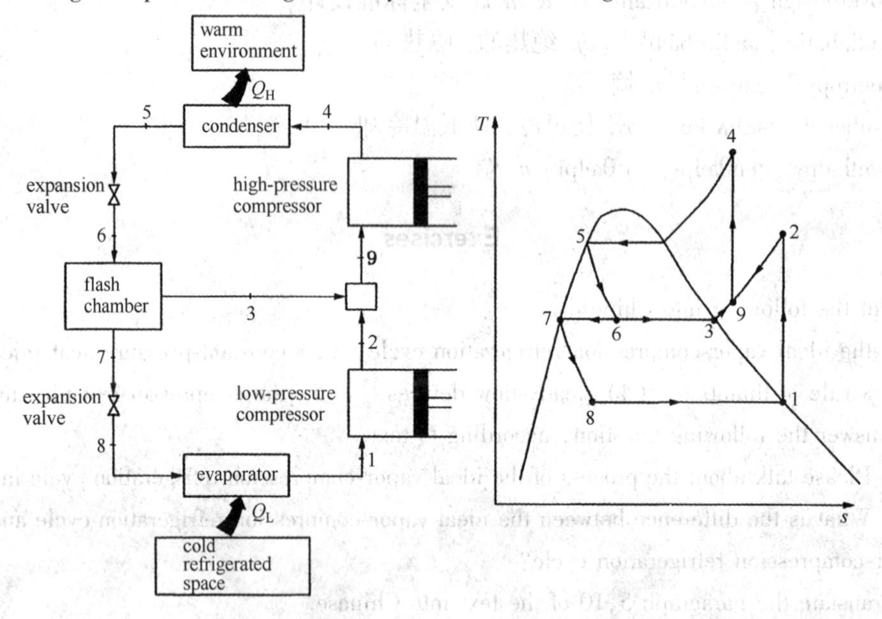

Fig.2.11 A two-stage compression refrigeration system with a flash chamber

In this system, the liquid refrigerant expands in the first expansion valve to the flash chamber pressure, which is the same as the compressor interstage pressure. Part of the liquid vaporizes during this process. This saturated vapor (state 3) is mixed with the superheated vapor from the low-pressure compressor (state 2), and the mixture enters the high-pressure compressor at state 9. This is, in essence, a regeneration process. The saturated liquid (state 7) expands through the second expansion valve into the evaporator, where it picks up heat from the refrigerated space.

The compression process in this system resembles a two-stage compression with intercooling, and the compressor work decreases. Care should be exercised in the interpretations of the areas on the $T-s$ diagram in this case since the mass flow rates are different in different parts of the cycle.

Cascade Refrigeration Systems

Some industrial applications require moderately low temperatures, and the temperature range they involve may be too large for a single vapor-compression refrigeration cycle to be practical. A large temperature range also means a large pressure range in the cycle and a poor performance for a reciprocating compressor. One way of dealing with such situations is to perform the refrigeration process in stages, that is, to have two or more refrigeration cycles that operate in series. Such refrigeration cycles are called cascade refrigeration cycles.

The two cycles of a two-stage cascade refrigeration cycle are connected through the heat exchanger in the middle, which serves as the evaporator for the topping cycle and the condenser for the bottoming cycle. In the cascade system, the refrigerants in both cycles are assumed to be the same. This is not necessary, however, since there is no mixing taking place in the heat exchanger. Therefore, refrigerants with more desirable characteristics can be used in each cycle. In this case, there would be a separate saturation dome for each fluid, and the $T-s$ diagram for one of the cycles would be different. Also, in actual cascade refrigeration systems, the two cycles would overlap somewhat since a temperature difference between the two fluids is needed for any heat transfer to take place.

It is evident that the compressor work decreases and the amount of heat absorbed from the refrigerated space increases as a result of cascading. Therefore, cascading improves the COP of a refrigeration system. Some refrigeration systems use three or four stages of cascading.

Words and Expressions

1. innovative [ˈinəuveitiv] *adj.* 创新的，革新(主义)的
2. modification [ˌmɔdifiˈkeiʃən] *n.* 更改，修改，修正
3. refinement [riˈfainmənt] *n.* 精致，(言谈，举止等的)文雅，精巧
4. flash chamber 闪蒸室
5. multistage [ˈmʌltisteidʒ] *adj.* 多级的
6. interstage [ˌintə(ː)ˈsteidʒ] *adj.* 级间的，级际的
7. intercooling [ˌintə(ː)ˈkuː liŋ] 中间冷却
8. interpretation [inˌtəːpriˈteiʃən] *n.* 解释，阐明，口译，通译
9. cascade [kæsˈkeid] *n.* 层叠
10. reciprocating [riˈsiprəkeitiŋ] 往复的，来回的，交替的，互换的，摆动的
11. overlap[ˈəuvəˈlæp] *v.* (与……)交迭

Exercises

1. Put the following into Chinese.

(1) household refrigerator (2) A two-stage compression refrigeration system

(3) interstage pressure (4) cascade refrigeration cycles (5) a separate saturation dome

2. Answer the following question, according to text.

(1) What is the disadvantages of the simple vapor-compression refrigeration cycle?.

(2) Which questions can cascade refrigeration cycles deal with?

3. Translate the paragraph 3, 5 of the text into Chinese.

Unit 13 Absorption Refrigeration

Ferdinand Carre, a Frenchman, invented the absorption system and took out a United States patent in 1860. The first use of the system in the United States was probably made by the Confederate States during the Civil War after the supply of natural ice had been cut off from the North.

Absorption refrigeration becomes economically attractive when there is a source of inexpensive thermal energy at a temperature of 100 to 250℃. Some examples of inexpensive thermal energy sources include geothermal energy, solar energy, and waste heat from cogeneration or process steam plants, and even natural gas when it is available at a relatively low price.

Absorption refrigeration systems involve the absorption of a refrigerant by a transport medium. The most widely used absorption refrigeration system is the ammonia-water system, where ammonia (NH_3) serves as the refrigerant and water (H_2O) as the transport medium. Other absorption refrigeration systems include water-lithium bromide and water-lithium chloride systems, where water serves as the refrigerant. The latter two systems are limited to applications such as air-conditioning where the minimum temperature is above the freezing point of water.

To understand the basic principles involved in absorption refrigeration, weshow the NH_3-H_2O system in Fig.2.12.You will immediately notice from the figure that this system looks very much like the vapor-compression system, except that the compressor has been replaced by a complex absorption mechanism consisting of an absorber, a pump, a generator, a regenerator, a valve, and a rectifier. Once the pressure of NH_3 is raised by the components in the box, it is cooled and condensed in the condenser by rejecting heat to the surroundings, is throttled to the evaporator pressure, and picks up heat from the refrigerated space as it flows through the evaporator. So, there is nothing new there. Here is what happens in the box:

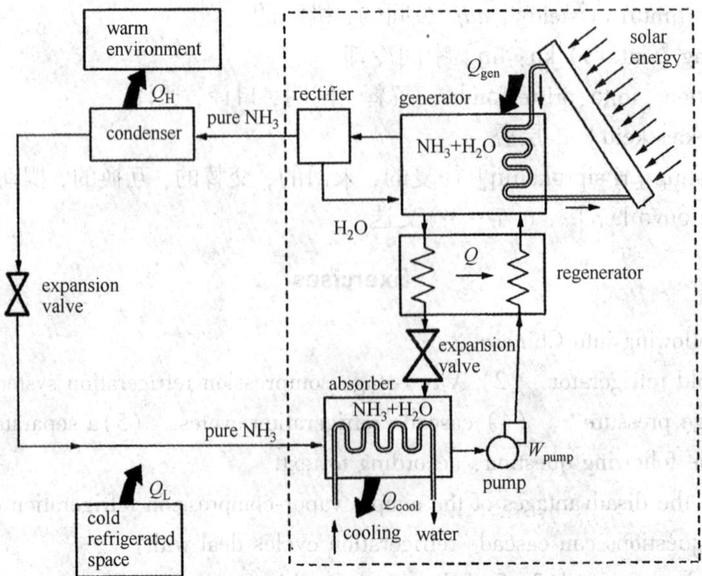

Fig.2.12 Ammonia absorption refrigeration cycle

Ammonia vapor leaves the evaporator and enters the absorber, where it dissolves and reacts with water to form $NH_3 \cdot H_2O$. This is an exothermic reaction; thus heat is released during this process. The amount of NH_3 that can be dissolved in H_2O is inversely proportional to the temperature. Therefore, it is necessary to cool the absorber to maintain its temperature as low as possible, hence to maximize the amount of NH_3 dissolved in water. The liquid $NH_3 + H_2O$ solution, which is rich in NH_3, is then pumped to the generator. Heat is transferred to the solution from a source to vaporize some of the solution. The vapor, which is rich in NH_3, passes through a rectifier, which separates the water and returns it to the generator. The high-pressure pure NH_3 vapor then continues its journey through the rest of the cycle. The hot $NH_3 + H_2O$ solution, which is weak in NH_3, then passes through a regenerator, where it transfers some heat to the rich solution leaving the pump, and is throttled the absorber pressure.

The COP of absorption refrigeration systems is defined as:

$$COP_R = \frac{Q_L}{Q_{gen} + W_{pump,in}} \tag{2-14}$$

The maximum COP of an absorption refrigeration system is determined by assuming that the entire cycle is totally reversible (i. e. , the cycle involves no irreversibilities and any heat transfer is through a differential temperature difference).

Compared with vapor-compression systems, the absorption refrigeration systems are much more expensive than the vapor-compression refrigeration systems. They are more complex and occupy more space, they are much less efficient thus requiring much larger cooling towers to reject the waste heat, and they are more difficult to service since they are less common. Therefore, absorption refrigeration systems should be considered only when the unit cost of thermal energy is low and is projected to remain low relative to electricity. Absorption refrigeration systems are primarily used in large commercial and industrial installations.

Absorption refrigeration systems have two major advantages compared with vapor-compression systems: (1) A refrigerant is compressed by a complex absorption mechanism instead of a vapor-compressor. (2) Absorption refrigeration systems are often classified as heat-driven systems. Because most of the operating cost of the absorption cycle is associated with providing the heat that drives off the vapor from the high-pressure liquid. Indeed there is a requirement for some work in the absorption cycle to drive pump, but the mount of work for a given quantity of refrigeration is minor compared with that needed in the vapor-compression cycle.

Words and Expressions

1. geothermal [ˌdʒi(ː)əuˈθəməl] *adj.* 地热的，地温的，地热(或地温)产生的
2. cogeneration [ˌkəudʒenəˈreiʃən] *n.* 利用余热发电
3. natural gas 天然气
4. lithium bromide 溴化锂
5. lithium chloride 氯化锂

6. rectifier[ˈrektifaiə] *n.* 精馏器(吸收式制冷机的一个外部冷却式换热器)
7. generator [ˈdʒenəreitə] *n.* 发电机，发生器
8. regenerator [riˈdʒenəreitə] *n.* 中间换热器，再生器
9. absorber [əbˈsɔːbə] *n.* 吸收器，吸收体
10. exothermic [ˌeksəuˈθɜːmik] *adj.* 发热的，放出热量的
11. cooling tower 冷却塔

Exercises

1. Put the following into Chinese.

(1) process steam plants　　(2) the freezing point of water

(3) heat-driven systems　　(4) a differential temperature difference

2. Answer the following question, according to text.

(1) Which absorption refrigeration systems are used widely?

(2) What are the similarities and differences between absorption refrigeration systems and vapor-compression refrigeration systems?

3. Translate the paragraph 3,5,8 of the text into Chinese.

Unit 14 Industrial Ventilation

The purpose of ventilation is to maintain in the building a prescribed condition and cleanliness of the air (in other words, the temperature, air velocity and concentrations). This task is resolved as follows: The vitiated air is removed from the building (extract ventilation), whilst in its place clean air is introduced, often specially treated (inflow ventilation).

In essence there are also heat transfer and mass transfer between the incoming air and the air already within the building. If owing to excessive internal heat production, the temperature of the air in the building tends to exceed the specified norms, cooler air is then remains at the norm. If harmful gases or vapors are released, their concentration is held within specified limits by dilution with the clean incoming air.

More often than not mass and heat transfer take place simultaneously. For instance, the production of convective heat is very often accompanied by the releases of gases and highly dispersed dust.

Ventilation can be affected by fans (mechanical ventilation) or by the difference between the densities of the columns of internal and external air, and also by the action of wind (natural ventilation).

If air is introduced into a building, some excess pressure is set up in it. In the steady state this pressure will be such that the total quantity of air leaving the building through specially provided vents, or through random crocks in the external surfaces, is equal to that which is introduced. A similar phenomenon will occur with the extract of air from the building. Here a negative pressure (rarefaction) is set up in the building, and in consequence air will be sucked in through gaps from outside and from adjacent rooms to take the place of the extracted.

In certain cases this air has an unfavorable effect. For instance, if cold outdoor air enters a building in which much water vapor is produced it would create mist on mixing with the internal hot and moist air. If the inflow from outside or from adjacent rooms satisfies the hygienic requirements, it can be used to replace general mechanical ventilation by natural ventilation.

Ventilation is essentially the science of the control of air change in building.

By extracting the air from areas with high concentrations of impurity, one considerably reduces the quantity of air needed for ventilation. For instance, in iron foundries the concentration of carbon monoxide in upper levels can be $0.04g/m^3$, whereas in the work area it should not exceed the permissible norm $0.02g/m^3$. This stratification of the concentration is maintained by a supply of fresh air near the floor, and by extracting the vitiated air from high level. If the fresh air were supplied near the ceiling, in descending it would disturb the stratification and mix with the vitiated air, and with the same air change the concentration of CO in the work area would be $0.03g/m^3$. To obtain a concentration of $0.02g/m^3$ one would have to increase the quantity of ventilation air by a factor of about 1.5. Thus the question of the estimated quantity of ventilating air is directly, related to the question of the arrangements for ventilation.

To calculate and design local ventilation, it is necessary to know the properties of the jet, the laws governing the variation of its velocity, temperature and concentration and the geometric dimensions of the jet. To obtain the hygienically prescribed parameters of the air at the workplace, one needs to know the initial parameters of the air and then find the forms of nozzles to produce a jet which would satisfy these requirements.

Words and Expressions

1. ventilation [venti'leiʃən] *n.* 通风，流通空气
2. concentration['kɔnsen'treiʃən] *n.* 集中，集合，专心，浓缩，浓度
3. vitiated ['viʃieitid] *adj.* 有害的
4. whilst [wailst] *conj.* 时时，同时
5. specified norms 设定的标准
6. dilution [dai'lju:ʃən] *n.* 稀释，稀释法，冲淡物
7. more often than not 时常
8. rarefaction [ˌrɛəri'fækʃən] *n.* 变稀薄，稀薄
9. hygienic [hai'dʒi:nik] *adj.* 卫生学的，卫生的
10. stratification [ˌstrætifi'keiʃən] *n.* 层化，成层，阶层的形成
11. nozzle ['nɔzl] *n.* 管口，喷嘴

Exercises

1. Put the following into Chinese.

(1) extract ventilation (2) dispersed dust (3) natural ventilation

(4) carbon monoxide

2. Answer the following question, according to text.

(1) What is the purpose of ventilation?

(2) What is ventilation essentially?

3. Translate the paragraph 2,3,4 of the text into Chinese.

Unit 15 Comfort Ventilation

Effective comfort ventilation is based on the principle that air must be delivered directly to the work zone with sufficient air motion and at a low enough temperature to cool the worker by convection and evaporation. In most cases, the objective is to provide tolerable working conditions rather than complete comfort. In each case, the methods described for comfort ventilation are based on the assumption that exhaust ventilation, radiation shielding, equipment insulation, and possible changes in process design have been fully used to minimize the heat loads. In additions, supply air must not flow on or at hot equipment or hoods nor through the layers of hot ceiling air before reaching the work zone. In the former case, the ventilation disturbs the capture velocity of the hood or thermal rise from the hot equipment and distributes hot and/or contaminated air into the workroom. In the latter case, huge volumes of hot and possibly contaminated air are entrained and brought down to the work zone. Because of the large volumes of ventilation required for industrial plants, heat conservation and recovery should be use. In some cases, it is possible to provide unheated, or partially heated make up air to the building. Rotary, regenerative + heat exchangers recover up to 80% of the heat represented by the difference between the exhaust and the outdoor air temperatures. Reductions of 5 to 10℃ in the heating requirement for most industrial systems should be routine. While most of the heat conservation and recovery methods outlined in this section apply to heating, the saving possibility of air-conditioning systems is equally impressive. Heat conservation and recovery should be incorporated in preliminary planning for an industrial plant. Some methods are:

(1) In the original design of the building, process, and equipment, provide insulation and heat shields to minimize heat loads. Vapor proofing and reduction of glass area may be required. Changes in process design may be required to keep the building heat loads within reasonable bounds. Review the exhaust needs for hoods and process and keep those to a practical, safe minimum.

(2) Design the supply air systems for efficient distribution by delivering the air directly to the work zones; by mixing the supply with hot building air in the winter by using recirculated air within the requirements of winter makeup; and by bringing unheated or partially heated air to hoods or process whenever possible.

(3) Design the system to achieve highest efficiency and lowest residence time for contaminants. In the design, consider that it is psychologically sound and good practice to permit workers to adjust and modify the air patterns to which they are exposed: people desire direct personal control over their working environment.

(4) Conserve exhaust air by using it; e. g. , office exhaust can be directed, first to work areas, then to locker rooms of process areas, and finally to the outside. Clean. heated air can be used from motor or generator rooms after it has been used for cooling the equipment. Similarly, the cooling systems for many large motors and generators have been arranged to discharge into the building in the winter to provide heat and to the outside in summer to avoid heat loads.

(5) Operate the system for economy. Shut the systems down at night or weekends whenever possible, and operate the makeup air in balance with the needs of operating process equipment and hoods. Keep supply air temperatures at the minimum for heating and the maximum for cooling, consistent with the needs of process and employee comfort. Keep the building in balance so that uncomfortable drafts do not require excessive heating.

Words and Expressions

1. exhaust [igˈzɔːst] *vi.* 排气;*n.* 排气，排气装置
2. insulation[ˌinsjuˈleiʃən] *n.* 绝缘
3. hood [hud] *n.* 通风橱，套，遮光罩，挡板
4. entrain [inˈtrein] *vt.* 夹带，产生，导致
5. unheated [ʌnˈhiːtid] *adj.* 未加热的，不热的
6. routine [ruːˈtiːn] *n.* 例行公事，常规，日常事务，程序
7. contaminant [kənˈtæminənt] *n.* 致污物，污染物

Exercises

1. Put the following into Chinese.

(1) comfort ventilation (2) equipment insulation (3) work zone (4) heat load

2. Answer the following question, according to text.

(1) What is comfort ventilation based on?

(2) How many methods are mentioned in the text for an industrial plant to incorporate heat conservation and recovery?

3. Translate the sentence 1, 2, 3, 4 of the paragraph 1 into Chinese.

PART Ⅲ RESEARCH PAPER AND WRITING OF PYROLOGY

Unit 1 Solar Energy

We have always used the energy of the sun as far back as humans have existed on this planet. As far back as 5,000 years ago, people "worshipped" the sun. Ra, the sun-god, who was considered the first king of Egypt. In Mesopotamia, the sun-god Shamash was a major deity and was equated with justice. In Greece there were two sun deities, Apollo and Helios. The influence of the sun also appears in other religions.

We know today, that the sun is simply our nearest star. Without it, life would not exist on our planet. We use the sun's energy every day in many different ways.

When we hang laundry outside to dry in the sun, we are using the sun's heat to do work—drying our clothes.

Plants use the sun's light to make food. Animals eat plants for food. And decaying plants hundreds of millions of years ago produced the coal, oil and natural gas that we use today. So, fossil fuels is actually sunlight stored millions and millions of years ago.

Indirectly, the sun or other stars are responsible forall our energy. Even nuclear energy comes from a star because the uranium atoms used in nuclear energy were created in the fury of a nova—a star exploding.

Let's look at ways in which we can use the sun's energy.

Solar Hot Water

In the 1890s solar water heaters were being used all over the United States. They proved to be a big improvement over wood and coal-burning stoves. Artificial gas made from coal was available to heat water, but it cost 10 times the price we pay for natural gas today. And electricity was even more expensive if you even had any in your town!

Many homes used solar water heaters. In 1897, 30 percent of the homes in Pasadena, just east of Los Angeles, were equipped with solar water heaters. As mechanical improvements were made, solar systems were used in Arizona, Florida and many other sunny parts of the United States.

By 1920, ten of thousands of solar water heaters had been sold. By then, however, large deposits of oil and natural gas were discovered in the westernUnited States. As these low cost fuels became available, solar water systems began to be replaced with heaters burning fossil fuels.

Today, solar water heaters are making a comeback. There are more than half a million of them in California alone! They heat water for use inside homes and businesses. They also heat swimming pools like in the picture.

Panels on the roof of a building, like that one in Fig.3.1, contain water pipes. When the sun hits the panels and the pipes, the sunlight warms them. That warmed water can then be used in a swimming pool.

Fig.3.1 Solar water heater

Solar Thermal Electricity

Solar energy can also be used to make electricity.

Some solar power plants, like the one in Fig.3.2 which is in California's Mojave Desert, use a highly curved mirror called a parabolic trough to focus the sunlight on a pipe running down a central point above the curve of the mirror. The mirror focuses the sunlight to strike the pipe, and it gets so hot that it can boil water into steam. That steam can then be used to turn a turbine to make electricity.

In California's Mojave Desert, there are huge rows of solar mirrors arranged in what's called "solar thermal power plants" that use this idea to make electricity for more than 350,000 homes. The problem with solar energy is that it works only when the sun is shining. So, on cloudy days and at night, the power plants can't create energy. Some solar plants, are a "hybrid" technology. During the daytime they use the sun. At night and on cloudy days they burn natural gas to boil the water so they can continue to make electricity.

Another form of solar power plants to make electricity is called a Central Tower Power Plant, like Fig.3.3—the Solar Two Project.

Fig.3.2 Solar power plant

Fig.3.3 Central tower power plant

Sunlight is reflected off 1,800 mirrors circling the tall tower. The mirrors are called heliostats and move and turn to face the sun all day long.

The light is reflected back to the top of the tower in the center of the circle where a fluid is turned very hot by the sun's rays. That fluid can be used to boil water to make steam to turn a turbine and a generator.

This experimental power plant is called Solar Ⅱ. It was re-built in California's desert using newer technologies than when it was first built in the early 1980s. Solar Ⅱ will use the sunlight to change heat into mechanical energy in the turbine.

The power plant will make enough electricity to power about 10,000 homes. Scientists say larger central tower power plants can make electricity for 100,000 to 200,000 homes.

Solar Cells or Photovoltaic Energy

We can also change the sunlight directly to electricity using solar cells.

Solar cells are also called photovoltaic cells-or PV cells for short-and can be found on many small appliances, like calculators, and even on spacecraft. They were first developed in the 1950s for use onU. S. space satellites. They are made of silicon, a special type of melted sand.

Shown in Fig. 3. 4, when sunlight strikes the solar cell, electrons (circles) are knocked loose. They move toward the treated front surface. An electron imbalance is created between the front and back. When the two surfaces are joined by a connector, like a wire, a current of electricity occurs between the negative and positive sides.

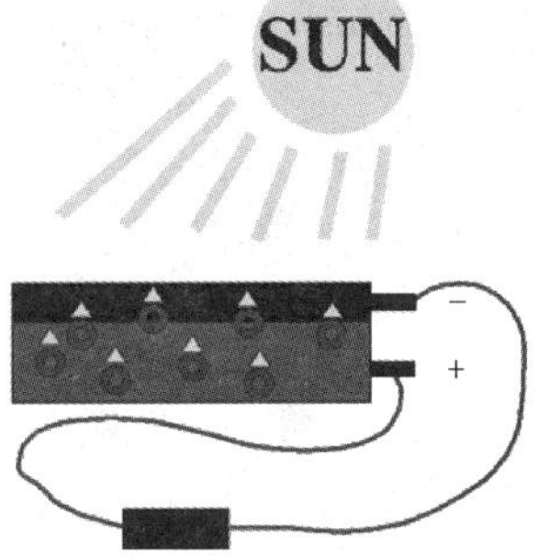

Fig.3.4 Solar cell operating principle

These individual solar cells are arranged together in a PV module and the modules are grouped together in an array. Some of the arrays are set on special tracking devices to follow sunlight all day long.

The electrical energy from solar cells can then be used directly. It can be used in a home for lights and appliances. It can be used in a business. Solar energy can be stored in batteries to light a roadside billboard at night. Or the energy can be stored in a battery for an emergency roadside cellular telephone when no telephone wires are around. Some experimental cars also use PV cells. They convert sunlight directly into energy to power electric motors on the car. But when most of us think of solar energy, we think of satellites in outer space.

Words and Expressions

1. Shamash [ˈʃaːməs] *n.* 沙玛什(巴比伦和亚述神话中的太阳神,象征正义)
2. Apollo [əˈpɔləu] *n.* 阿波罗(太阳神), 美男子
3. Helios [ˈhiːliɔs]*n.* 太阳神
4. nova[ˈnəuve]*n.* 新星
5. curved mirror 曲面镜
6. parabolic[ˌpærəˈbɔlik] *adj.* 抛物线的, 抛物线状的
7. trough[ˈtrɔːf] *n.* 槽, 水槽
8. hybrid [ˈhaibrid] *n.* 混合物, 杂种
9. heliostat [ˈhiːliəustæt] *n.* 定日镜, 日光反射装置
10. photovoltaic [ˌfəutəuvɔlˈteiik]*adj.* 光电(效应)的; *n.* 光电伏打电池

Exercises

1. Put the following into Chinese.

(1) artificial gas (2) thermal electricity (3) all day long

(4) solar cell (5) in an array (6) cellular telephone

2. Answer the following question, according to text.

(1) Why do we say without sun, life would not exist on our planet?

(2) What is the "hybrid" technology?

(3) Narrate the operational principle of the solar cell.

3. Translate the paragraph 4, 13 and the last paragraph into Chinese.

Unit 2 Nuclear Energy-Fission and Fusion

Another major form of energy is nuclear energy, the energy that is trapped inside each atom. One of the laws of the universe is that matter and energy can't be created nor destroyed. But they can be changed in form.

Matter can be changed into energy. The world's most famous scientist, Albert Einstein, created the mathematical formula that explains this. It is: $E = mc^2$

This equation says: E [energy] equals m [mass] times c^2 [c stands for the velocity or the speed of light. c^2 means c times c, or the speed of light raised to the second power-or c-squared.].

Please note that some web browser software may not show an exponent (raising something to a power, a mathematical expression) on the Internet. Normally c-squared is shown with a smaller "2" placed above and to the right of the c.

Scientists used Einstein's famous equation as the key to unlock atomic energy and also create atomic bombs.

The ancient Greeks said the smallest part of nature is an atom. But they did not know 2,000 years ago about nature's even smaller parts.

As we learned, atoms are made up of smaller particles (As shown in Fig.3.5) —a nucleus of protons and neutrons, surrounded by electrons which swirl around the nucleus much like the earth revolves around the sun.

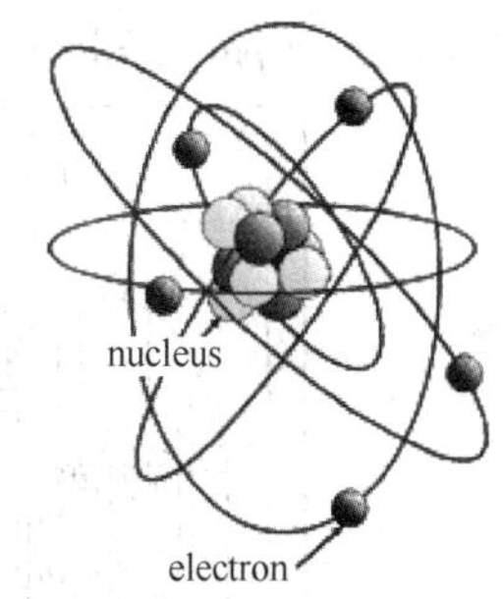

Fig.3.5 The structure of atom

Nuclear Fission

An atom's nucleus can be split apart. When this is done, a tremendous amount of energy is released. The energy is both heat and light energy. Einstein said that a very small amount of matter contains a verylarge amount of energy. This energy, when let out slowly, can be harnessed to generate electricity. When it is let out all at once, it can make a tremendous explosion in an atomic bomb.

Fig.3.6 Diablo Canyon nuclear plant

A nuclear power plant (As shown in Fig.3.6) uses uranium as a "fuel". Uranium is an element that is dug out of the ground many places around the world. It is processed into tiny pellets that are loaded into very long rods that are put into the power plant's reactor.

The word fission means to split apart. Inside the reactor of an atomic power plant, uranium atoms are split apart in a controlled chain reaction.

In a chain reaction, particles released by the splitting of the atom go off and strike other uranium atoms splitting those. Those particles given off split still other atoms in a chain reaction, in nuclear power plants, control rods are used to keep the splitting regulated so it doesn't go too fast.

If the reaction is not controlled, you could have an atomic bomb. But in atomic bombs, almost pure pieces of the element Uranium-235 or Plutonium, of a precise mass and shape, must be brought together and held together, with great force. These conditions are not present in a nuclear reactor.

The reaction also creates radioactive material. This material could hurt people if released, so it is kept in a solid form. The very strong concrete dome in the picture is designed to keep this material inside if an accident happens.

This chain reaction gives off heat energy. This heat energy is used to boil water in the core of the reactor. So, instead of burning a fuel, nuclear power plants use the chain reaction of atoms splitting to change the energy of atoms into heat energy.

This water from around the nuclear core is sent to another section of the power plant. Here, in the heat exchanger, it heats another set of pipes filled with water to make steam. The steam in this second set of pipes turns a turbine to generate electricity. Below is a cross section of the inside of a typical nuclear power plant, as shown in Fig.3.7.

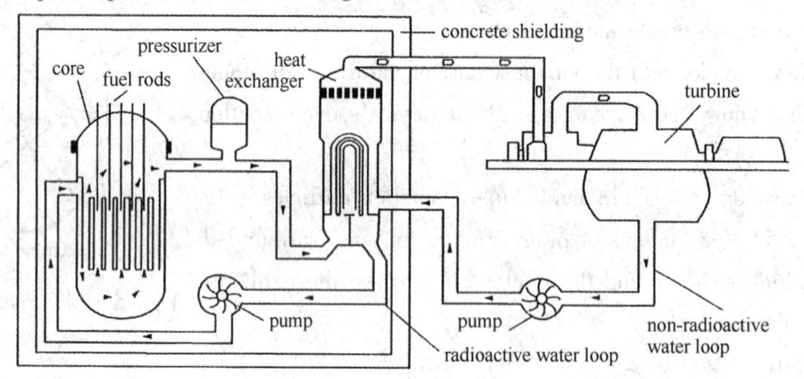

Fig.3.7 Power plant drawing courtesy nuclear institute

Nuclear Fusion

Another form of nuclear energy is called fusion. Fusion means joining smaller nuclei (the plural of nucleus) to make a larger nucleus. The sun uses nuclear fusion of hydrogen atoms into helium atoms. This gives off heat and light and other radiation.

In the picture to the right, two types of hydrogen atoms, deuterium and tritium, combine to make a helium atom and an extra particle called a neutron.

Also given off in this fusion reaction is energy! Thanks to the University ofCalifornia, Berkeley for the picture.

Scientists have been working on controlling nuclear fusion for a long time, trying to make a fusion reactor to produce electricity. But they have been having trouble learning how to control the reaction in a contained space.

What's better about nuclear fusion is that it creates less radioactive material than fission, and its supply of fuel can last longer than the sun.

Words and Expressions

1. fission ['fiʃən] *n.* 【原】裂变
2. fusion ['fju:ʒən] *n.* 【原】聚变
3. nucleus ['nju:kliəs] *n.* 核心，核子，原子核
4. uranium [juə'reidiəm] *n.* 铀
5. radioactive ['reidiəu'æktiv] *adj.* 放射性的，有放射能的
6. give off 发出(蒸汽、光等)
7. plutonium [plu:'təuniəm] *n.* 【化】钚
8. hydrogen ['haidrəudʒən] *n.* 氢
9. deuterium [dju:'tiəriəm]*n.* 【化】氘
10. tritium ['tritiəm] *n.* 【化】氚(氢的放射性同位素)
11. neutron ['nju:trɔn] *n.* 中子

Exercises

1. Put the following into Chinese.

(1) atomic energy (2) atomic bomb (3) nuclear fission
(4) nuclear fusion (5) nuclear power plant

2. Answer the following question, according to text.

(1) Can you tell about the definition of the fission and the fusion?
(2) What's the Albert Einstein's famous equation? Point out the signification of every sign?

3. Translate the paragraph 8, 10, 11, 12, 16 of the text into Chinese.

Unit 3 Renewable Energy vs. Fossil Fuels

Some of the energy we can use is called renewable energy. Renewable energy sources include solar energy, which comes from the sun and can be turned into electricity and heat. Wind, geothermal energy from inside the earth, biomass from plants, and hydropower and ocean energy from water are also renewable energy sources. These types of energy are constantly being renewed or restored.

The kinetic energy of the wind can be changed into other forms of energy, either mechanical energy or electrical energy.

People use the geothermally heated hot water in swimming pools and in health spas. Or, the hot water from below the ground can warm buildings for growing plants. Hot water or steam from below ground can also be used to make electricity in a geothermal power plant.

Some ofbiomass is just stuff lying around—dead trees, tree branches, yard clippings, left-over crops, wood chips, and bark and sawdust from lumber mills. It can even include used tires and livestock manure. This stuff can be used to produce electricity, heat, compost material or fuels.

When it rains in hills and mountains, the water becomes streams and rivers that run down to the ocean. The moving or falling water, which has kinetic energy, can be used tooperate gristmill and watermill or make electricity.

There are three basic ways to tap the ocean for its energy. One is wave energy. Kinetic energy exists in the moving waves of the ocean. That energy can be used to power a turbine or cylinder which can turn a generator. Most wave-energy systems are very small. But, they can be used to power a warning buoy or a small light house. Another is tidal energy. When tides come into the shore, they can be trapped in reservoirs behind dams. Then when the tide drops, the water behind the dam can be let out just like in a regular hydroelectric power plant. We can also use ocean thermal energy. Power plants can be built that use this difference in temperature to make energy. A difference of at least 38 degrees Fahrenheit is needed between the warmer surface water and the colder deep ocean water.

However, we get most of our energy from nonrenewable energy sources, which include the fossil fuels—oil, natural gas, and coal. Another nonrenewable energy source is the element uranium, whose atoms we split (through a process called nuclear fission) to create heat and ultimately electricity.

Fossil fuels take millions of years to make. We are using up the fuels that were made more than 300 million years ago before the time of the dinosaurs. Fossil fuels are not renewable, they can't be made again. Once they are gone, they're gone.

Emissions from cars fueled by gasoline and factories and other facilities that burn oil affect the atmosphere. Foul air results in so-called greenhouse gases. About 81% of all U. S. greenhouse gases are carbon dioxide emissions from energy-related sources.

In contrast, clean energy sources can be harnessed to produce electricity, process heat, fuel and valuable chemicals with less impact on the environment. Renewable energy resource development will result in new jobs for people and less oil we have to buy from foreign countries. According to the federal government, America spent $ 109 billion to import oil in 2000. If they fully develop self-renewing resources, they will keep the money at home to help the economy.

Continued research has made renewable energy more affordable today than 25 years ago. But there are also drawbacks to renewable energy development.

For example, solar thermal energy involving the collection of solar rays through collectors (often times huge mirrors) need large tracts of land as a collection site. This impacts the natural habitat, meaning the plants and animals that live there. The environment is also impacted when the buildings, roads, transmission lines and transformers are built. The fluid most often used with solar thermal electric generation is very toxic and spills can happen.

Solar or PV cells use the same technologies as the production of silicon chips for computers. The manufacturing process uses toxic chemicals. Toxic chemicals are also used in making batteries to store solar electricity through the night and on cloudy days. Manufacturing this equipment has environmental impacts.

Also, even if we wanted to switch to solar energy right away, we still have a big problem. All the solar production facilities in the entire world only make enough solar cells to produce about 350 megawatts, about enough for a city of 300,000 people, that's a drop in the bucket compared to our needs. And the cost of producing that much electricity would be about four times more expensive than a regular natural gas-fired power plant.

So, even though the renewable power plant doesn't release air pollution or use precious fossil fuels, it still has an impact on the environment.

Wind power development too, has its downside, mostly involving land use. The average wind farm requires 17 acres of land to produce one megawatt of electricity, about enough electricity for 750 to 1,000 homes. However, farms and cattle grazing can use the same land under the wind turbines.

Wind farms could cause erosion in desert areas. Most often, winds farms affect the natural view because they tend to be located on or just below ridge lines. Bird deaths also occur due to collisions with wind turbines and associated wires. This issue is the subject of on-going research.

Producing geothermal electricity from the earth's crust tends to be localized. That means facilities have to be built where geothermal energy is abundant. In the course of geothermal production, steam coming from the ground becomes very caustic at times, causing pipes to corrode and fall apart. Geothermal power plants sometimes cost a little bit more than a gas-fired power plant because they have to include the cost to drill.

Environmental concerns are associated with dams to produce hydroelectric power. People are displaced and prime farmland and forests are lost in the flooded areas above dams. Downstream, dams change the chemical, physical and biological characteristics of the river and land.

Unlike fossil fuel, which dirties the atmosphere, renewable energy has less impact on the environment. Renewable energy production has some drawbacks, mainly associated with the use of large of tracts of land that affects animal habitats and outdoor scenery.

Words and Expressions

1. renewable energy 可再生能源
2. biomass [ˈbaiəuˌmæs] *n.* 生物质
3. spa [spaː] *n.* 矿泉,温泉场
4. bark [baːk] *n.* 树皮, 吠声
5. sawdust [ˈsɔːdʌst] *n.* 据屑, 锯末
6. manure [məˈnjuə] *n.* 肥料
7. buoy [bɔi] *n.* 浮标, 浮筒, 救生圈
8. Fahrenheit [ˈfærənhait] *n.* 华氏温度计;*adj.* 华氏温度计的
9. megawatt [ˈmegəwɔt] *n.* 兆瓦特
10. a drop in the bucket 微不足道
11. ridge line 脊线

Exercises

1. Put the following into Chinese.

(1) kinetic energy (2) wave energy (3) tidal energy

(4) nonrenewable energy (5) Once they are gone, they're gone (6) left over

2. Answer the following question, according to text.

(1) What are included in renewable energy sources?

(2) Narrate the three basic ways to utilize ocean energy.

(3) Are the fossil fuels renewable? Why?

3. Translate the paragraph 1, 12, 18 into Chinese.

Unit 4 Technological Application of High Temperature Air Combustion in Diluted/Rich Conditions

Combustion processes are controlled by fluid-dynamic, thermodynamic and composition variables. Among them the temperature is the most representative one in characterizing the whole process. It is common to refer to combustion as occurring at low and high etc. temperature. This is a loose way of classifying such complex process, which usually needs further specification with respect to the stage of the process to which it refers. There is further complexity when there is more than one temperature relevant to the combustion process.

Other definition has been used during the last years. Mild Combustion is a subset of the HTAC or HCOT dominion, as it is shown in some papers of the aforementioned Symposia. Mild Combustion process is relevant if the inlet temperature and the temperature increase are respectively slightly higher and lower than the autoignition one.

This means that the whole process evolves in a quite narrow temperature range, which could be placed in an intermediate region between the very fast kinetics of the oxidative undiluted conditions and the relatively slow kinetics linked to the low temperature autoignition regimes.

"A combustion process in a given reactor is named 'mild' when the autoignition temperature of the reactants is lower than the inlet temperature of the main reactant flow and higher than the maximum allowable temperature increase in the reactor."

The reactor parameters which are relevant here are the reactant composition (fuel, oxidant and dilute), the pressure (the definition applies only to nearly isobaric reactor) and the minimum residence time of the reactants. The autoignition temperature referred to the example reported here is that pertaining both to a stoichiometric homogeneous fuel/air mixture at the reactor pressure and to an autoignition time equal to the minimum residence time in the reactor.

The maximum temperature increase (T) is the difference between the maximum temperature, which can occur in the reactor, and the maximum temperature of the inlet reactant. T varies according to the reactor type. For instance, if the reactants are not premixed, the maximum temperature is the adiabatic flame temperature of the system referred to the stoichiometric condition. In fact, these conditions may also occur when the feeding ratio is different from the stoichiometric. On the other hand, the maximum temperature of a premixed stream is fixed by the feeding ratio between fuel/oxidant/dilute. It is also noteworthy that the maximum temperature is related to the maximum oxidation level, which may be different from both the equilibrium and real temperature gained in the reactor. For instance a $CH_4/O_2/N_2$ system evolving in rich, diluted conditions may attain different temperatures according to different kinetic routes followed in the chemical process that, in turn, do not necessary lead to the maximum oxidation level. It pertains to a Well-Stirred Reactor temperature increase (T_r) plotted versus the inlet temperature for $C/O = 1$ and $O_2/N_2 = 0.06$ and a fixed residence time of 1sec (solid line). Details related to this analysis are fully described in a previous work. The dotted and the dashed lines are the values T computed for the equilibrium (T_{eq}), and the theoretical adiabatic flame temperature (T_{af}) of the system. In the case here considered T_{af} is always higher than T_r because in the rich diluted condition the main oxidation product is CO.

The shape of solid line is due to different product distributions obtained by changing inlet temperature. The three arrows related to the curve refer to different outlet compositions according to the different oxidation channels.

The usefulness of the Mild Combustion definition may be appreciated by some considerations in relation to the map reported in Fig obtained for the same chemical system presented before, i. e. $CH_4/O_2/N_2$ with 0. 1/0. 05/0. 85 molar fractions. It defines all possible inlet temperature (abscissa) and temperature increase (ordinate) for a residence time of 1sec and atmospheric pressure. In this case the autoignition temperature is 1000K according to an evaluation based on a numerical computation.

The map is divided in three regions of interest by the straight lines intercepting the autoignition temperature on both axes. These regions are named Feedback, High Temperature and Mild Combustion respectively. According to the definition above reported (i. e. $T<T_{in}<T_{ign}$) the Mild Combustion is placed in the lower-right quadrant. The other two combustion modes are placed in the upper part of the map where the condition $T>T_{ign}$ is satisfied.

The meaning of the Mild Combustion in comparison to the other two fields is quite straight. In fact it differs from the other two regimes because in the Mild Combustion the process cannot be sustained without the preheating of the reactants. On the contrary, Feedback and High Temperature Combustion satisfy the necessary condition for which a combustion process may occur, namely the heat release is sufficient to sustain the minimum temperature required the process evolution. .

Related definitions to Mild Combustion are High Temperature Reactant Combustion, Flameless Combustion and Colorless Combustion. Flameless or Colorless. Combustion are definitions which pertain to properties of the reactors rather than to inlet condition. They refer to the outstanding characteristic that no visible emission is detectable in the oxidation regions. The first one is more restrictive in the sense that some other properties are also linked to the colorless features. It is hard to say at the moment whether these definitions are coincident with Mild Combustion. However, all these processes rely on highly preheated and diluted systems, but it is difficult to fix the quantitative extent to which the different definitions apply. If the definitions are shown to be the same, the Mild Combustion could be the criterion to satisfy for the identification of Flameless or Colorless regime but, at the moment, the only suitable choice in the use of the definitions pertains to the emphasis which is needed on the inlet conditions and for the precision required in the text.

Words and Expressions

1. parameter [pə'ræmitə] *n.* 参数
2. definition [ˌdefi'niʃən] *n.* 定义，解说，精确度，(轮廓影像等的)清晰度
3. characterize ['kæriktəraiz] *vt.* 表现……的特色
4. mild [maild] *adj.* 温和的，淡味的，轻微的，适度的，不含有害物质的

5. subset [ˈsʌbset] *n.* 【数】子集
6. allowable [əˈlauəbl] *adj.* 允许的，正当的，可承认的
7. dilute [daiˈljuːt] *v.* 冲淡，变淡，变弱，稀释；*adj.* 淡的，弱的，稀释的
8. equilibrium [ˌiːkwiˈlibriəm] *n.* 平衡，平静，均衡，保持平衡的能力，沉着，安静

Exercises

1. Put the following into Chinese.

(1) mild combustion (2) adiabatic flame temperature (3) the maximum temperature of

(4) different kinetic routes (5) flameless combustion (6) colorless combustion

2. Answer the following question, according to text.

(1) How combustion process are controlled?

(2) What is Mild Combustion?

(3) What parameters are relevant to the reactor?

3. Translate the paragraph into Chinese.

(1) Combustion processes are controlled by fluid-dynamic, thermodynamic and composition variables. Among them the temperature is the most representative one in characterizing the whole process.

(2) The meaning of the Mild Combustion in comparison to the other two fields is quite straight. In fact it differs from the other two regimes because in the Mild Combustion the process cannot be sustained without the preheating of the reactants.

(3) It is hard to say at the moment whether these definitions are coincident with Mild Combustion. However, all these processes rely on highly preheated and diluted systems, but it is difficult to fix the quantitative extent to which the different definitions apply.

Unit 5 Advanced Control of Walking-Beam Reheating Furnace

Walking-beam reheating furnace is an important device with lots of energy consumption in steel plants. It is important to improve the heating quality of slab and reduce the energy consumption as much as possible. In current system, the emphasis is often put on the heating quality while the energy saving is seldom taken into account. Because of high nonlinearity large time delay large time-constant and various uncertain factors, the modeling and reliable control of a reheating furnace are always challenging problems. During the past decade, the challenge has attracted considerable attention and a significant progress has been made in the furnace control. To obtain the requirement of temperature tracing accuracy and develop energy saving technique in the reheating furnace system, advanced control strategies are needed. In this paper, a model-based predictive control method, multivariable constrained control (MCC), was used. It treated the furnace as a multi-input-multi-output system and used the predefined temperature trajectories as a priori. The simulation results show that the strategy is very satisfactory.

Multivariable advanced process control (APC) techniques have been gradually accepted in process industries, which pursue high product quality and lower cost, and have increased environmental responsibility. APC is usually model based and regulates the processes with optimization algorithms. In APC, it can usually improve the control of a process significantly by minimizing the variation in the controlled process variables, dealing with interaction among process variables and automatically controlling set-points of PID loops, which would be normally regulated by operators. In this paper, a new APC-multivariable constraint control (MCC) is proposed for the advanced control of the reheating furnace. Its kernel algorithm is derived from model predictive control (MPC), which has been greatly developed in recent 20 years, and has almost become the accepted standard for constrained multivariable control problems in the process industries. Here at each sampling time, starting at the current state, an open loop optimal control problem is solved over a finite horizon. At the next time step the computation is repeated starting from the new state and over a shifted horizon, leading to a moving horizon policy.

The results (MVs) from dynamic optimal block will be sent back to runtime database (platform), and another control cycle will begin from step 1 again.

The general model of the reheating furnace is a six-inputs-six-outputs system. Constraints in the system include the limit of furnace temperature, the limit of temperature change in a computation period, the limit of fuel flux, the limit of change ratio of air to gas, etc.

Setting of optimization index

The optimization index can be set according to different control requirements. If the temperature tracing accuracy is demanded for high-qualified productions, Q is to be set larger. If the energy saving demand is more important, the Q can be set smaller where R, S are set larger accordingly, which results the smoother fuel flux.

In the optimization index, the temperature tracing error, the fuel flux value and its changes are taken into account explicitly. All these values can be regulated to reach specific requirements by modifying corresponding weight matrix. Taking the fuel flux change into account in the performance index is useful to make the fuel flux changes smoother, for large fluctuation of fuel flux is harmful to the combustion in the furnace. With the optimization index, it is obviously to see that the accuracy of temperature tracing and the values of fuel flux can be adjusted simultaneously in a systematical way.

In spite of handling input and output constraints explicitly, the MCC controller make decoupling implemented naturally for its multivariable handling abilities. From a practical viewpoint, these features can great enhance the performance of the reheating furnace.

Numerical results

The simulation result of the constructed control system will be shown to demonstrate the effect of the proposed control strategy. In this system the initialization is realized by reading recorded data from industry site. Hence, the initial condition is the same as the real system. MCC controller and traditional PID controller are used and the simulation results are figured. The parameters of PID controller are the same as the real plant. Because setting of the optimization index of MCC is for temperature tracing accuracy requirement, so the matrix Q is most important and matrix R, S are set for slight changes requirement of fuel flux.

The simulation covers 5h. Here only show one simulation result, the temperature traced results of preheating and heating zone, with no change of the controller parameters of PID and MCC. The two zones include four manipulated variables and four controlled variables, which are tightly coupled.

The fluctuation of the fuel flux has significant impact on the value of residual oxygen content, an important index for combustion analysis in the furnace. Lower value of residual oxygen content means more sufficiently fuel combustion and more energy saving. Because MCC can regulate the fuel flux by change corresponding weight matrix in the on-line optimization, it is also an ideal energy saving technique.

The fuel flux provided by loops with MCC controller is much smoother than that provided by loops without MCC controller. The fuel flux provided by loops with only PID controller has a large oscillation when set-points of furnace temperature change.

The results in some experiments show that the tracing accuracy is largely improved by using MCC controller. The reason is that MCC is based on multivariable control strategy, where the decoupling work is implemented in a systematical and implicit way. That is why advanced control becomes more and more important in complex multivariable industry process.

Conclusions

(1) A better temperature tracing accuracy is obtained compared with the result of a traditional controller.

(2) The implemented strategy is a kind of optimization method and some energy saving techniques can be designed accordingly.

(3) The fluctuation of fuel flux is taken into consideration in the optimization index explicitly,

so gas combustion in the furnace will be more exhaustive. Energy saving means more profits to steel plants.

Words and Expressions

1. consumption [kən'sʌmpʃən] *n.* 消费，消费量
2. regulate ['regjuleit] *vt.* 管制，控制，调节，校准
3. optimization [ˌɔptimai'zeiʃən] *n.* 最佳化，最优化
4. simulation [ˌsimju'leiʃən] *n.* 仿真，假装模拟
5. significant [sig'nifikənt] *adj.* 有意义的，重大的，重要的
6. multivariable [ˌmʌlti'veəriəbl] *adj.*【统】【数】多变量的，多元的
7. traditional [trə'diʃən(ə)l] *adj.* 传统的，惯例的，口传的，传说的
8. performance [pə'fɔːməns] *n.* 履行，执行，成绩，性能，表演，演奏
9. initialization [iˌniʃəlai'zeiʃən] *n.* 设定初值，初始化
10. demonstrate ['demənstreit] *vt.* 示范，证明，论证
11. trajectory ['trædʒiktəri, trə'dʒekətəri] *n.*【物】(射线的)轨道，弹道，轨线，曲线

Exercises

1. Put the following into Chinese.

(1) walking-beam (2) the fuel flux value (3) the constructed control system

(4) lots of energy consumption

2. Answer the following question, according to text.

(1) what is an important index for combustion analysis in the furnace?

(2) What is the importance of the Walking-beam reheating furnace ?

3. Translate the paragraph into Chinese.

(1) The results (MVs) from dynamic optimal block will be sent back to runtime database (platform), and another control cycle will begin from step 1 again.

(2) The implemented strategy is a kind of optimization method and some energy saving techniques can be designed accordingly.

(3) The results in some experiments show that the tracing accuracy is largely improved by using MCC controller. The reason is that MCC is based on multivariable control strategy, where the decoupling work is implemented in a systematical and implicit way. That is why advanced control becomes more and more important in complex multivariable industry process.

Unit 6 Heat Exchangers

Introduction

Heat exchangers are one of the most common pieces of equipment found in all plants. Their purpose is very simple: to heat a cold stream using a hot one or cool a hot stream by using a cold one. They are usually a much cheaper alternative to heat or cool fluids as opposed to electric heaters or coolers because they often use fluids that are already present in the plant. For example, to heat a cold stream, steam condensate from a boiler may be rerouted to pass through a heat exchanger and heat up a cold fluid. As for a cooler, water from a nearby stream could be piped into cool off a hot fluid. A plate exchanger (from VED Engineering [1]) is shown in Fig 3.8. They are available with several variations, such as different plate lengths, gaskets, etc. It is comparable to the plate cooler used in the lab here at UNB.

Fig 3.8 Industial plate exchanger (VED Engeering)

Unlike the heat exchangers found in our department lab, heat exchangers can be immense pieces of equipment. Seen below in Fig.3.9 is a set of industrial shell-and-tube heat exchangers, courtesy of Bronswerk Heat Transfer BV [2]. Exchangers of this size and durability can handle very high pressures (1000 Bar), corrosive environments, and temperatures ranging from−196 up to more than 800 °C. Fig 3.10 gives an interesting view of the inside of a shell and tube exchanger, where the individual tubes are installed to contain tube-side flow.

Fig.3.9 Industial shell and tube heat exchanger(Bronswerk heat transfer BV)

Fig.3.10 Inside view of shell and tube exchanger (Bronswerk heat transfer BV)

The purpose of this experiment was to study the operation of three types of heat exchangers in an attempt to simulate a larger setup. The working fluid in use, Dowtherm SR-1, will be sent through the heat exchanger network and the effectiveness of each exchanger under different flow conditions will be found.

From the temperatures at various points on the apparatus, experimental values for the overall heat transfer coefficient can be calculated. These values can be compared to theoretical calculations based on flow rates and the properties of the equipment. The calculated overall heat transfer coefficients will be used to predict how the actual setup will respond under different conditions. This experiment is an excellent illustration of many important topics covered in ChE3304, since it deals very closely with the determination of the overall heat transfer coefficient in a system of heating and cooling between two fluids.

The main objectives for this experiment is to see which heat exchanger, shell-and-tube or steam condenser, should be used and in what configuration, either co-current or counter-current. This decision will be based on the minimum steam and/or boiler condensate requirements as well as cooling water requirements.

Objectives:

- To determine the overall heat transfer coefficients and effectiveness for each heat exchanger.
- To determine the most feasible setup for a heat exchanger network

Theory

In this experiment, the effectiveness of each heat exchanger is at the heart of the decision of which one would be the best choice for a given industrial application. The effectiveness of a heat exchanger is the efficiency with which the exchanger transfers heat from the hot stream to the cold stream, or in the case of a cooler, how well a cold stream can cool a hot stream. It is of great importance because it dictates the requirements of the hot or cold utilities, which can sometimes be quite costly.

The effectiveness depends on the rate at which heat is transferred from one fluid to the other. The heat is transferred by convection from the hot fluid to the tube wall. It travels through the tube wall by conduction, and then to the cold fluid by convection once again. The rate at which heat is transferred by convection depends on the heat transfer coefficient, h, of that fluid. The greater the heat transfer coefficient, the easier the heat moves from one fluid to the other. For conduction, the ease with which heat is conducted is shown by the thermal conductivity, k, of the material.

Heat exchanger effectiveness is also dependent on the nature of the material the exchanger is built out of. Some materials conduct heat better than others, so it is essential to have a material with a high thermal conductivity in order to have good transfer of heat from one fluid to the other. The effectiveness of the heat exchanger is also dependent on the fluids flowing through the tubes. The specific heats and the flow rates have a significant effect on the rate of heat transferred, so it is also of interest.

The effectiveness can be found experimentally by comparing the maximum and minimum mC_p terms, as is shown below:

$$\varepsilon=\frac{q}{q_{max}}=\frac{C_H(T_{H,in}-T_{H,out})}{C_{min}(T_{H,in}-T_{C,in})}=\frac{C_C(T_{C,out}-T_{C,in})}{C_{min}(T_{H,in}-T_{C,in})} \tag{3-1}$$

Where $C_H=m_H C_{pH}, C_C=m_C C_{pC}$

The C_{min} value represents whichever value is the lesser quantity between C_H or C_C. Theoretically, the NTU (number of transfer units) method is used:

$$\varepsilon=\frac{1-\exp\left[\frac{UPL}{C_{min}}(1+C)\right]}{1+C} \tag{3-2}$$

Where $C = \frac{C_{min}}{C_{max}} \rightarrow$ flow thermal capacity.

In this equation, U is the experimental overall heat transfer coefficient, P is the inside perimeter of the pipe, and L is the length of the pipe. P and L are given in the specifications and C_{min} and C can be found graphically (C_P values) and by recording the flow rates of the fluids, so these values can be used to find the experimental overall heat transfer coefficient once the effectiveness is calculated.

The experimental overall heat transfer coefficient is calculated by using the log-mean temperature difference, $\ddot{A}\ T_{LM}$. This is a value that is calculated based on hot and cold flow temperatures entering and leaving the heat exchanger. The log mean temperature difference represents the appropriate mean temperature between hot and cold streams for a heat exchanger [3]. Once the LMTD has been calculated, it can be used to find the experimental overall heat transfer coefficient.

To observe the accuracy of the experimental results, theoretical calculations must also be done. To do this, the *Nusselt* Number (*Nu*) is used. The *Nusselt* number is a measure of the heat transfer of a fluid in terms of its heat transfer coefficient, thermal conductivity, and the geometry of the pipe or object. *Nusselt* numbers are needed for the both fluids, shell and tube side, because the overall heat transfer requires both inside and outside heat transfer coefficients to calculate U. Different equations for the *Nusselt* number are used for different conditions (e. g. Reynolds number, *Prandl* number) and can be found in texts [3].

[1] VED Engineering. http://www. vedengineering. com/

[2] *Bronswerk* Heat Transfer BV. http://www. bronswerk. nl/ *Stationsweg* 22, 3862 CG *Nijkerk P. O. Box* 92, 3860 *AB Nijkerk*, The Netherlands

[3] *Mills*, *A. F.* Heat Transfer-Second Edition. *Prentice Hall. Upper Saddle River*, *NJ.* 1999

Words and Expressions

1. perimeter [pə'rimitə] *n.* 【数】周长，周界
2. condensate [kɔn'denseit] *n.* 冷凝物，冷凝液
3. plate exchanger 板式换热器
4. gasket ['gæskit] *n.* 束帆索，垫圈，衬垫
5. shell-and-tube 管壳式
6. durability [ˌdjuərə'biliti] *n.* 耐久性，耐用性，持久性经久，耐久力
7. corrosive [kə'rəusiv] *adj.* 腐蚀的，蚀坏的，腐蚀性的
8. coefficient [kəui'fiʃənt] *n.* 系数
9. steam condenser 蒸汽冷凝器
10. configuration [kənˌfigju 'reiʃən] *n.* 构造，结构，配置，外形
11. convection [kən'vekʃən] *n.* 传送，对流

Exercises

1. Put the following into Chinese.

(1) heat exchanger (2) shell-and-tube heat exchanger (3) plate exchanger
(4) steam condenser (5) co-current (6) counter-current

2. Answer the following question, according to text.

(1) How many types of heat exchangers are listed in this text? Can you list them?

(2) What does heat exchanger effectiveness mainly depend on?

3. Translate the paragraph 8,9,10 of the text into Chinese.

Unit 7 Introduction to Fluid Mechanics

Historical development of fluid mechanics

The science of fluid mechanics began with the need to control water for irrigation and navigation purposes in ancient China, Egypt, Mesopotamia, and India. Although these civilizations understood the nature of channel flow, there is no evidence that any quantitative relationships had been developed to guide them in their work. It was not until 250 B. C. that Archimedes discovered and recorded the principles of hydrostatics and buoyancy. In spite of the fact that the empirical understanding of hydrodynamics continued to improve with the development of fluid machinery, better sailing vessels, and more intricate canal systems, the fundamental principles of classical hydrodynamics were not founded until the seventeenth and eighteenth centuries. Newton, Daniel Bernoulli, and Leonard Euler made the greatest contributions to the founding of these principles.

In the nineteenth century, two schools of thought arose in the treatment of fluid mechanics, one dealing with the theoretical and the other with practical aspects of fluid flow. Classical hydrodynamics, though a fascinating subject that appealed to mathematicians, was not applicable to many practical problems because the theory was based on inviscid fluids. The practicing engineers at that time needed design procedures that involved the flow of viscous fluids; consequently, they developed empirical equations that were usable but narrow in scope. Thus, on the one hand, the mathematicians and physicists developed theories that in many cases could not be used by the engineers, and on the other hand, engineers used empirical equations that could not be used outside the limited range of application from which they were derived. In a sense, these two schools of thought have persisted to the present day, resulting in the mathematical field of hydrodynamics and the practical science of hydraulics.

Near the beginning of the twentieth century, however, it was necessary to merge the general approach of the physicists and mathematicians with the experimental approach of the engineer to bring about significant advances in the understanding of flow processes. Osborne Reynolds' paper in 1883 on turbulence and later papers on the basic equations of liquid motion contributed immeasurably to the development of fluid mechanics. After the turn of the century, in 1904, Ludwig Prandtl proposed the concept of the boundary layer. In this short, convincing paper Prandtl, at a stroke, provided an essential link between ideal and real fluid motion for fluids with a small viscosity and provided the basis for much of modem fluid mechanics.

Scope and significance of fluid mechanics

Fluid mechanics, as the name indicates, is that branch of applied mechanics which is concerned with the static and dynamics of liquid and gases. Dynamics, the study of motion of matter, may be divided into two parts-dynamics of rigid bodies and dynamics of non-rigid bodies. The latter is usually further divided into two general classifications-elasticity (solid elastic body) and fluid mechanics.

The subject of fluid mechanics can be subdivided into two broad categories: hydrodynamics and gas dynamics. Hydrodynamics deals primarily with the flow of fluids for which there is virtually no density change, such as liquid flow or the flow of gas at low speeds. Hydraulics, for example, the study of liquid flows in pipes or open channels, falls within this category. The study of fluid forces on bodies immersed in flowing liquids or in low-speed gas flows can also be classified as hydrodynamics.

Gas dynamics, on the other hand, deals with fluids that undergo significant density changes. High-speed gas flowing through a nozzle or over a body, the flow of chemically reacting gases, or the movement of a body through the low density air of the upper atmosphere falls within the general category of gas dynamics.

An area of fluid mechanics not classified as either hydrodynamics or gas dynamics is aerodynamics, which deals with the flow of air past aircraft or rockets, whether it is low-speed incompressible flow or high-speed compressible flow.

There are, however, two major aspects of fluid mechanics which differ from solid-body mechanics. The first is the nature and properties of the fluid itself, which are very different from those of a solid. The second is that, instead of dealing with individual bodies or elements of known mass, we are frequently concerned with the behavior of a continuous stream of fluid, without beginning or end.

Knowledge and understanding of the basic principles and concepts of fluid mechanics are essential in the analysis and design of any system in which a fluid is the working medium. Many applications of fluid mechanics make it one of the most vital and fundamental of all engineering and applied scientific studies. The flow of fluids in pipe and channels makes fluid mechanics importance to civil engineers. The study of fluid machinery such as pumps, fans, blowers, compressors turbines, heat exchangers, jet and rocket engines, and the like, makes fluid mechanics of importance to mechanical engineers. Lubrication is an area of considerable importance in fluid mechanics. The flow of air over objects, aerodynamics, is of fundamental interest to aeronautical and space engineers in the design of aircraft, missiles and rockets. In meteorology, hydrology and oceanography the study of fluids is basic since the atmosphere and the ocean are fluids. And today in modern engineering many new disciplines combine fluid mechanic with classical disciplines. For example, fluid mechanics and electromagnetic theory are studied together as magnetogasdynamics, in new types of energy conversion devices and in the study of stellar and ionospheric phenomena, magnetogasdynamics is vital.

On the contrary, the collapse of the Tacoma Narrows Bridge in U. S. A. is evidence of the possible consequences of neglecting the basic principles of fluid mechanics. On a memorable day in November 1940, Nature decided to teach us all a lesson. The wind could not even be considered strong on that day, but it happened to disturb the great Tacoma Narrows suspension bridge cyclically with a frequency close to the bridge's natural frequency of vibration. The entire bridge started to dance. Traffic was stopped; and an astonished public watched the bridge dance itself to pieces.

Words and Expressions

1. mechanics [mi'kæniks] *n.* 机械学，力学

2. hydrostatics [ˌhaidrəuˈstætiks] *n.* 流体静力学
3. buoyancy [ˈbɔiənsi] *n.* 浮性, 浮力, 轻快
4. hydrodynamics [ˈhaidrəudaiˈnæmiks] *n.* 流体力学, 水动力学
5. inviscid [inˈvisid] *adj.* 非黏滞性的;无韧性的;不能展延的
6. viscous [ˈviskəs] *adj.* 黏性的, 黏滞的, 胶黏的
7. hydraulics [ˈhaiˈdrɔːliks] *n.* 水力学
8. viscosity [visˈkɔsiti] *n.* 黏质, 黏性
9. rigid [ˈridʒid] *adj.* 刚硬的, 刚性的, 严格的
10. nozzle [ˈnɔzl] *n.* 管口, 喷嘴
11. aerodynamics [ˌεərəudaiˈnæmiks] *n.* 空气动力学, 气体力学
12. vibration [vaiˈbreiʃən] *n.* 振动, 颤动, 摇动, 摆动
13. turbine [ˈtəːbin] *n.* 涡轮
14. lubrication [ˌljuːbriˈkeiʃən] *n.* 润滑油
15. meteorology [ˌmiːtjəˈrɔlədʒi] *n.* 气象学, 气象状态
16. hydrology [haiˈdrɔlədʒi] *n.* 水文学, 水文地理学
17. oceanography [ˌəuʃiəˈnɔgrəfi] *n.* 海洋学
18. magnetogasdynamics [mægˈniːtəuˌgæsdaiˈnæmiks] *n.* 磁气体动力学

Exercises

1. Put the following into Chinese.

(1) fluid mechanics (2) inviscid fluids (3) viscous fluids
(4) empirical equation (5) gas dynamics (6) solid-body mechanics

2. Answer the following question, according to text.

(1) What purpose did fluid mechanics apply in ancient China, Egypt, Mesopotamia, and India?

(2) What are the major differences between fluid mechanics and solid-body mechanics?

3. Translate the paragraph 1, 4, 5, 9 of the text into Chinese.

Unit 8　Gas Turbine

Most gas-turbine engines (Fig. 3. 11) include at least a compressor, a combustion chamber, and a turbine. These are usually mounted as an integral unit and operate as a complete prime mover on a so-called open cycle where air is drawn in from the atmosphere and the products of combustion are finally discharged again to the atmosphere. Since successful operation depends on the integration of all components, it is important to consider the whole device, which is actually an internal-combustion engine, rather than the turbine alone.

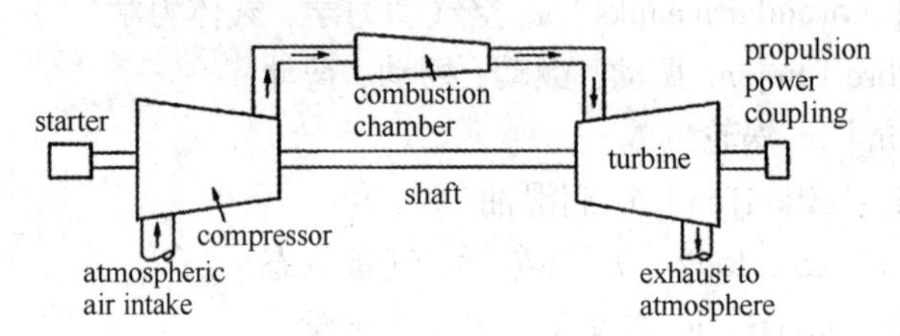

Fig. 3. 11　A gas turbine

Gas turbines were developed primarily for aircraft industry and for internal combustion engine supercharges, now they have played a unique role in electric utility systems. Because of their relatively low initial cost, compactness and quick starting (they can be on the line at rated load within 30 minutes), the plants are suitable for emergency service and "peaking" service during daily periods of high load demand.

Gas turbine development accelerated in the 1930s as a means of propulsion for jet aircraft. It was not until the early 1980s that the efficiency and reliability of gas turbines had progressed sufficiently to be widely adopted for stationary power applications. Gas turbines range in size from 30 kW (micro-turbines) to 250 MW (industrial frames). Industrial gas turbines have efficiencies approaching 40% and 60% for simple and combined cycles respectively.

Natural gas is the most suitable fuel for gas turbines, however, natural gas is expensive for power production and is not available at most sites. Heavy oils are less expensive and are more available for power generation, thus are used extensively in gas turbines. Preheating must be employed in order to reduce the viscosity for proper flow and injection. Fuel oil heating is usually accomplished by shell and tube heat exchangers served by steam supplied by an external source.

Gas turbine plants operate on the Brayton cycle. The operating principle of gas turbine is basically as follows: ambient air is drawn into a multistage compressor and is compressed to about 10 times atmospheric pressure. The compressed air then passes through the combustion chamber where the preheated fuel is injected and burned. The products of combustion enter the turbine and expand to approximately atmosphere pressure. The turbine section powers both the generator and compressor.

About two-thirds of the turbine output is used to drive the compressor while the remainder is for power generation. This arrangement is called the simple cycle gas turbine, which is characterized, by large exhaust energy loss. In general, the simple-cycle gas turbine has relatively low efficiency

(25% to 30%) as compared with coal-fired steam turbine system. The efficiency of simple cycle can be increased by installing a recuperator or waste heat boiler to the turbine's exhaust part. A recuperator captures waste heat in the turbine exhaust stream to preheat the compressor discharge air before it enters the combustion chamber. A waste heat boiler generates steam by capturing heat from the turbine exhaust. These boilers are known as heat recovery steam generators (HRSG). They can provide steam for heating or industrial processes, which is called cogeneration. High-pressure steam from these boilers can also generate power with steam turbines, which is called a combined cycle (steam and combustion turbine operation). Recuperators and HRSGs can increase the gas turbine's overall energy cycle efficiency up to 80%. The combined-cycle plant is shown in Fig 3.12.

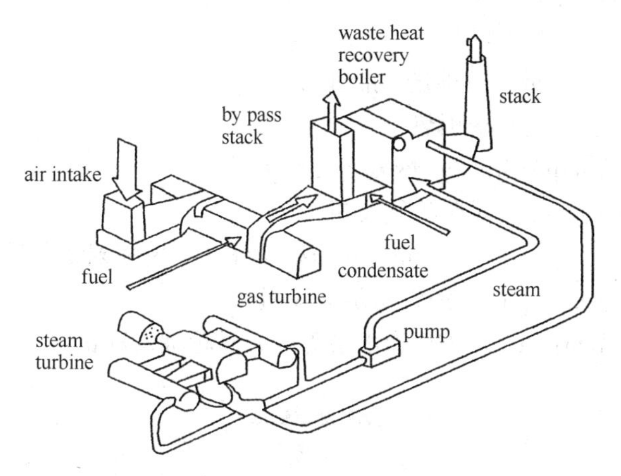

Fig. 3.12 A combined-cycle plant

It is an extension of the gas turbine plant, consisting of one or more gas turbine units generating electric power, with the hot exhaust gas discharged into waste heat recovery boilers. Steam is generated in the waste heat recovery boiler to serve a steam turbine generator, and the cooled exhaust gases are discharged to the atmosphere through a stack. Additional fuel may be burned in the waste heat recovery boiler, if required to supplement the heat in the gas turbine exhaust. The steam produced in the waste heat recovery boiler is expanded in the steam turbine and condensed in a surface type condenser. The condensate is returned to the boiler where it is preheated by the gas turbine exhaust.

There are four main types of gas turbine: turbojet, turbofan, turboprop, and turboshaft.

The turbojet and turbofan are both reaction engines which derive power from the reaction to the exhaust stream. The turboprop and turboshaft operate differently by using the exhaust stream to power an additional turbine which drives a propeller or output shaft. The original concept, the turbojet, is the simplest form of gas turbine and relies on the high velocity hot gas exhaust to provide the thrust. Its disadvantages today are its relatively high noise levels and fuel consumption.

In the turbofan or bypass' engine the partly compressed airflow is divided, some into a central part-the gas generator or core-and some into a surrounding casing-the bypass duct. The gas generator acts like a turbojet whilst the larger mass of bypass air is accelerated relatively slowly down the duct to provide 'cold stream' thrust. The cold and hot streams mix to give better propulsive efficiency, lower noise levels, and improved fuel consumption. In the high bypass ratio turbofan, as much as seven or eight times as much air bypasses the core as passes through it. It achieves around 75% of its thrust from the bypass air and is ideal for subsonic transport aircraft.

As its name implies, a turboprop uses a propeller to transmit the power it produces. The propeller is driven through a reduction gear by a shaft from a power turbine, using the gas energy which would provide the thrust in a turbojet.

Words and Expressions

1. supercharges [ˌsjuːpəˈtʃaːdʒ] *vt.* 对……增压
2. peaking [piːkiŋ] *n.* 剧烈增加
3. propulsion [prəˈpʌlʃən] *n.* 推进, 推进力
4. preheating [ˈpriːhiːtiŋ] *v.* 预热
5. recuperator [riˈkjuːpəreitə] *n.* 恢复者, 复原者
6. turbojet [ˈtəbəudʒet] *n.* 涡轮喷气飞机
7. turbofan [ˈtəːbəuˌfæn] *n.* 扇涡轮, 扇喷射(标准涡轮喷射机的改良型)
8. turboprop [ˈtəːbəuprɔp] *n.* 涡轮螺旋桨飞机
9. turboshaft [ˈtəːbəuʃæft] *n.* (驱动水泵等的) 涡轮轴发动机

Exercises

1. Put the following into Chinese.

(1) gas turbine (2) heavy oils (3) fuel oil (4) recovery boiler
(5) turbojet (6) turbofan (7) turboprop (8) turboshaft

2. Answer the following question, according to text.

(1) Why the cycle in a gas turbine is called open cycle?
(2) How can the efficiency of simple cycle in gas turbine be increased?
(3) Describe the operation process of a combined cycle (steam and combustion turbine operation).

3. Translate the paragraph 1, 2, 5, 6 into Chinese.

Unit 9 Integrated Gasification Combined Cycle

Integrated Gasification Combined Cycle (IGCC) is a process in which a low-value fuel such as coal, petroleum coke, orimulsion, biomass or municipal waste is converted to low heating value, high-hydrogen gas in a process called gasification. The gas is then used as the primary fuel for a gas turbine. IGCC can also be viewed as the two-stage combustion of a feedstock. First, the feedstock is partially combusted in a reactor or gasifier. Then the combustion is completed in the gas turbine.

IGCC Consists of Four Process

(1) Gasification

A feedstock can be gasified in several ways. The most common technique partially oxidizes the feedstock with pure oxygen inside a reactor. The carbon and hydrogen from the feedstock are converted into a mixture composed primarily of hydrogen and carbon monoxide. This mixture is commonly called synthetic gas, or syngas. Syngas has a heating value of 125 to 350 BTU/scf, which is three to eight times lower than that of natural gas.

(2) Syngas Cleanup

The syngas from the reactor must be cleaned before it can be used as a gas turbine fuel. The cleanup process typically involves removing sulphur compounds, ammonia, metals, alkalies, ash and particulates to meet the gas turbine's fuel gas specifications. To make IGCC more economically attractive, you can make marketable products such as methanol, ammonia, fertilizers and other chemicals from the compounds you remove from the syngas. This process often further reduces the hydrogen content and therefore the heating value of the syngas.

(3) Gas Turbine Combined Cycle

The cleaned syngas is combusted in the gas turbine.

(4) Cryogenic Air Separation

A cryogenic air separation unit provides pure oxygen to the gasification reactor, often using or supplemented with post-compression air bleed from the gas turbine.

The four process islands must be integrated to optimize the plant. Syngas, air, nitrogen and multi-pressure steam must be piped across the plant.

IGCC Benefits

Water consumption of an IGCC plant is approximately 30 percent lower than a conventional coal plant. Also, lime or limestone is not required for desulphurization. IGCC plants are highly competitive commercially, producing electricity at costs below that of conventional solid fuel plants. IGCC also offers benefits in terms of:

Output enhancement

Emissions reduction

Reliability, availability and maintenance (RAM) performance

(1) Output Enhancement

Because syngas has a low heating value compared to natural gas, significantly more fuel must be injected in an IGCC turbine than a natural gas turbine. Therefore, the mass flow-and thus the output power-of the gas turbine is much higher for an IGCC application. For the same reason, the gas turbine's output power is higher.

(2) Emission Reduction

IGCC SO_x, NO_x, and particle emissions are fractions of those of a conventional pulverized coal boiler power plant. As a consequence, meeting air emissions regulations and obtaining local and governmental environmental permits for an IGCC plant requires significantly less effort and time. IGCC enjoys recognition from environmental groups as the best environmental solution for power from coal.

To moderate NO_x emissions, steam, water, carbon dioxide and/or nitrogen can be injected into the gas turbine's combustor. Nitrogen is usually available from the cryogenic air separation unit, so it is convenient to use in the IGCC process. Fuel moisturization using low-level process heat is also used frequently. These techniques can achieve NO_x emissions that are similar to dry low NO_x(DLN) technology. The best available control technology reduces NO_x emissions to about 15 ppm currently, and is expected to reduce NO_x emissions to less than ten ppm in the near future.

During the IGCC process, harmful pollutants are removed from the synthetic gas before they reach the gas turbine, so exhaust gas cleanup is unnecessary.

IGCC technology facilitates the removal of a high percentage of mercury at low cost. IGCC gas turbines do not require expensive back-end flue gas mercury removal systems. Activated carbon bed filters in syngas and recycled water streams remove 90 to 95 percent of the mercury for only $20 to $30 USD per kW installed. The carbon bed has a lifetime of 12 to 18 months.

In IGCC plants, carbon can be removed from the syngas before combustion to create a high-hydrogen fuel, effectively eliminating carbon dioxide emissions. In conventional boiler plants, carbon is removed from the exhaust gas after combustion, which is much more expensive due to the larger gas volume from post-combustion cleanup (about 10 to 100).

(3) Reliability, Availability & Maintenance

The cost of electricity is affected by reliability, availability and maintenance (RAM) performance in a manner similar to plant cost and fuel efficiency. Generally, RAM performance is improved during the design stage by incorporating lessons learned from previous plants. Gas turbine suppliers can do some of this based on their experience in other fields, but actual IGCC experience is critical.

GE has more than 350,000 hours of directly relevant experience, including some turbines that have accumulated more than 30,000 hours of operation on synthetic gas (syngas). Based on this experience, they conclude that an IGCC plant can have the same RAM performance as a natural gas combined cycle plant.

The IGCC gas turbine must meet specific conditions to realize this high level of performance. For example, the higher hydrogen content in the syngas fuel which produces more water and the increased flow of syngas increase metal temperatures in the hot gas path. GE has developed a control system that mitigates this effect by lowering the firing temperature to keep the metal temperatures similar to those in natural gas turbines.

Words and Expressions

1. petroleum coke 石油焦
2. orimulsion *n.* 奥里乳化油
3. feedstock [ˈfiːdstɔk] *n.* 给料
4. monoxide [məˈnɔksaid] *n.* 一氧化物
5. syngas [ˈsiŋgæs] *n.* 合成气（指用煤等做原料制成的一氧化碳和氢的混合物）
6. ammonia [ˈæməunjə] *n.* 氨
7. methanol [ˈmeθənɔl] *n.* 甲醇
8. cryogenic [ˌkraiəuˈdʒenik] *adj.* 低温学的
9. fertilizer [ˈfəːtiˈlaizə] *n.* 肥料
10. moisturization *n.* 增加水分，润

Exercises

1. Put the following into Chinese.

(1) heating value (2) carbon monoxide (3) synthetic gas

(4) mass-flow (5) pulverized coal (6) back-end

2. Answer the following question, according to text.

(1) What is the IGCC?

(2) IGCC have some benefits, what are they?

3. Translate the paragraph 2, 7, 9 into Chinese.

Unit 10　Several Boilers Introduction

General Boiler Information

1 boiler horse power is about 42,000 BTUs of input. 1 pound of steam is about 1,200 BTUs of input fuel, and about 1,000 BTUs at the point of use, depending on the pressure of the steam. Low pressure steam is considered to be up to 15 psig; high is generally 100 psig and higher.

Superheat is a term that refers to higher temperature steam, a result of a second special steam heat exchanger in the boiler that allows steam pressure to increase, thereby taking on more BTUs (in excess of 500 psig is typical of superheat). Superheated steam is very dry steam.

Smaller boilers are generally rated in horse power; larger are generally rated in thousands of pounds of steam (500 hp and under will typically be rated in hp).

Typical boiler efficiency will be in the 75%～85% range; new highest efficiency boilers may be near 90%; newer quick heat up types of boilers with copper heat exchangers can be more efficient, especially at startup and part load than older, heavy mass cast iron boilers.

Fire Tube Boilers (Fig. 3.13)

In fire tube boilers, the combustion gases pass inside boiler tubes, and heat is transferred to water on the shell side. Scotch marine boilers are the most common type of industrial fire tube boiler. The Scotch marine boiler is an industry workhorse due to low initial cost, and advantages in efficiency and durability. Scotch marine boilers are typically cylindrical shells with horizontal tubes configured such that the exhaust gases pass through these tubes, transferring energy to boiler water on the shell side.

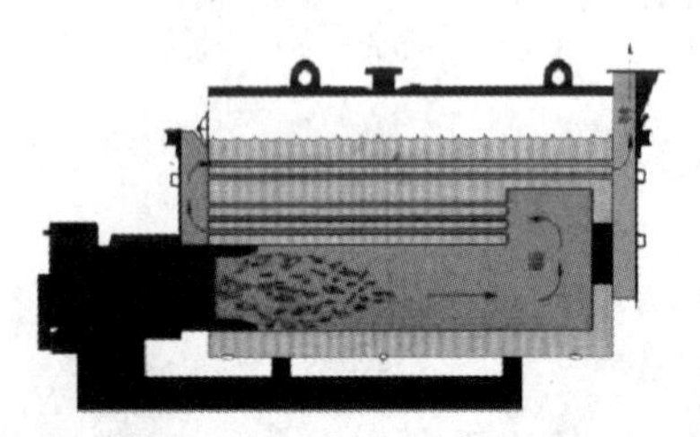

Fig. 3.13　Fire Tube Boilers

Scotch marine boilers contain relatively large amounts of water, which enables them to respond to load changes with relatively little change in pressure. However, since the boiler typically holds a large water mass, it requires more time to initiate steaming and more time to accommodate changes in steam pressure. Also, Scotch marine boilers generate steam on the shell side, which has a large surface area, limiting the amount of pressure they can generate. In general, Scotch marine boilers are not used where pressures above 300 psig are required. Today, the biggest fire tube boilers are over 1, 500 boiler horsepower.

Fire tube boilers are often characterized by their number of passes, referring to the number of times the combustion (or flue) gases flow the length of the pressure vessel as they transfer heat to the water. Each pass sends the flue gases through the tubes in the opposite direction. To make another pass, the gases turn 180 degrees and pass back through the shell. The turnaround zones can be either dry-back or water-back. In dry-back designs, the turnaround area is refractory lined. In water-back designs, this turnaround zone is water-cooled, eliminating the need for the refractory lining.

Water Tube Boilers (Fig. 3.14)

In water tube boilers, boiler water passes through the tubes while the exhaust gases remain in the shell side, passing over the tube surfaces. Since tubes can typically withstand higher internal pressure than the large chamber shell in a fire tube, water tube boilers are used where high steam pressures (as high as 3,000 psig) are required.

Fig. 3.14 Water Tube Boilers

Water tube boilers are also capable of high efficiencies and can generate saturated or superheated steam. The ability of water tube boilers to generate superheated steam makes these boilers particularly attractive in applications that require dry, high-pressure, high-energy steam, including steam turbine power generation.

The performance characteristics of water tube boilers make them highly favorable in process industries, including chemical manufacturing, pulp and paper manufacturing, and refining. Although fire tube boilers account for the majority of boiler sales in terms of units, water-tube boilers account for the majority of boiler capacity.

Steam Generators (Fig. 3.15)

Steam generators are like boilers in that they are fired by gas and produce steam, but they are unlike boilers in that they do not have large pressure vessels and are made of light-weight materials. The fact that they do not have pressure vessels means that in most locations they do not require a boiler operator (always confirm with local codes). This can be a substantial savings when there is no other reason to have an operator other than the local code requires it for a large pressure vessel. The fact that they are made out of light weight materials means they perform well at part loads and respond quickly to changes in loads. This greatly increases part load operating efficiency.

Fig. 3.15 Steam Generators

Compact and Modular Boilers (Fig. 3.16)

Modern materials, controls and the pursuit of even higher energy efficiency and reduced emissions are leading to boilers that are smaller in physical size, have cleaner emissions and produce dryer steam. Materials are critical because old cast iron boilers relied on mass to prevent them from thermal shocks that could split the boiler apart. New metals reduce mass which improves thermal transfer and can handle the thermal stress of going from cold water to steam in seconds.

Space is money, especially in new construction. Boilers of similar output capacity made smaller to reduce their space requirements can result in overall lower first cost of equipment plus space.

Boilers have a certain efficiency curve that tends to result in the boiler having the highest efficiency at full-fire. As the boiler is 'turned-down' to run at less than 100% capacity, efficiency typically drops. To counter this situation, a modular boiler bank of 3 or more boilers with a programmed

Fig. 3.16 Compact and Modular Boilers

controller can improve operational efficiency over a single boiler operating in a turn-down mode, and provides a certain amount of redundancy for back-up.

Tubeless and Condensing Boilers (Fig. 3.17)

High energy prices along with improved material and combustion technology is resulting in a new generation of high efficiency gas boilers. Traditional boilers are designed to prevent condensation because it is corrosive to boiler components and the wide variations in temperatures cause problems with thermal shock. However, without condensation, boilers can not be higher than about 85% efficient. Boilers that are designed for condensation and use advanced controls to squeeze every possible BTU from the combustion process are able to achieve efficiencies in the high 90's. There is a first-cost premium, but when energy prices are high, paybacks are more acceptable.

Fig. 3.17 Tubeless and Condensing Boilers

"Tubeless" Boilers use tubing coils instead of rigid tubes. "Direct Contact" water heaters have no tubes, tubing or coils; they have heat transfer media such as spheres or cylinders and allow flue gases to come in direct contact with the water.

Words and Expressions

1. superheat [ˌsjuːpəˈhiːt] *n.* 【物】过热
2. combustion [kəmˈbʌstʃən] *n.* 燃烧
3. startup *n.* 启动
4. horizontal [ˌhɔriˈzɔntl] *adj.* 地平线的，水平的
5. cylindrical [siˈlindrikəl] *adj.* 圆柱的
6. accommodate [əˈkɔmədeit] *vt.* 供应，供给，使适应，调节，向……提供，容纳，调和
7. psig abbr. pounds per square inch, gauge 磅/平方英寸（表 压）
8. BTU abbr. British Thermal Unit 英国热量单位
9. vessel [ˈvesl] *n.* 容器，器皿，脉管，导管
10. refractory [riˈfræktəri] *adj.* 难控制的，难熔的

11. saturate['sætʃəreit] v. 使饱和，浸透，使充满
12. refine [ri'fain] vt. 精炼，精制
13. emission [i'miʃən] n. (光、热等的) 散发，发射，喷射
14. condensation [kɔnden'seiʃən] n. 浓缩，冷凝

Exercises

1. Put the following into Chinese.
(1) boiler horse power (2) Fire Tube Boilers (3) exhaust gases
(4) Water Tube Boilers (5) Tubeless and Condensing Boilers
2. Answer the following question, according text.
(1) Which several kinds of boilers are listed in this text?
(2) In fire tube boilers, what's function of the combustion gases?
3. Translate the paragraph 5, 8, 9, 10, 11 of the text into Chinese.

Unit 11 Boiler Water Treatment

Origin of the Problem

The most common source of corrosion in boiler systems is dissolved gas: oxygen, carbon dioxide and ammonia. Of these, oxygen is the most aggressive. The importance of eliminating oxygen as a source of pitting and iron deposition cannot be over-emphasized. Even small concentrations of this gas can cause serious corrosion problems.

Makeup water introduces appreciable amounts of oxygen into the system. Oxygen can also enter the feed water system from the condensate return system. Possible return line sources are direct air-leakage on the suction side of pumps, systems under vacuum, the breathing action of closed condensate receiving tanks, open condensate receiving tanks and leakage of nondeaerated water used for condensate pump seal and/or quench water. With all of these sources, good housekeeping is an essential part of the preventive program.

One of the most serious aspects of oxygen corrosion is that it occurs as pitting. This type of corrosion can produce failures even though only a relatively small amount of metal has been lost and the overall corrosion rate is relatively low. The degree of oxygen attack depends on the concentration of dissolved oxygen, the pH and the temperature of the water.

The influence of temperature on the corrosivity of dissolved oxygen is particularly important in closed heaters and economizers where the water temperature increases rapidly. Elevated temperature in itself does not cause corrosion. Small concentrations of oxygen at elevated temperatures do cause severe problems. This temperature rise provides the driving force that accelerates the reaction so that even small quantities of dissolved oxygen can cause serious corrosion.

The Corrosion Process

Localized attack on metal can result in a forced shutdown. The prevention of a forced shutdown is the true aim of corrosion control.

Because boiler systems are constructed primarily of carbon steel and the heat transfer medium is water, the potential for corrosion is high. Iron is carried into the boiler in various forms of chemical composition and physical state. Most of the iron found in the boiler enters as iron oxide or hydroxide. Any soluble iron in the feed water is converted to the insoluble hydroxide when exposed to the high alkalinity and temperature in the boiler.

These iron compounds are divided roughly into two types, red iron oxide (Fe_2O_3) and black magnetic oxide (Fe_3O_4). The red oxide (hematite) is formed under oxidizing conditions that exist, for example, in the condensate system or in a boiler that is out of service. The black oxides (magnetite) are formed under reducing conditions that typically exist in an operating boiler.

External Treatment

External treatment, as the term is applied to water prepared for use as boiler feed water, usually refers to the chemical and mechanical treatment of the water source. The goal is to improve the quality of this source prior to its use as boiler feed water, external to the operating boiler itself. Such external treatment normally includes:

(1) Clarification
(2) Filtration
(3) Softening
(4) Dealkalization
(5) Demineralization
(6) Deaeration
(7) Heating

Any or all of these approaches can be used in feed water or boiler water preparation.

Internal Treatment

Even after the best and most appropriate external treatment of the water source, boiler feed water (including return condensate) still contains impurities that could adversely affect boiler operation. Internal boiler water treatment is then applied to minimize the potential problems and to avoid any catastrophic failure, regardless of external treatment malfunction.

Feed Water Preparation

The basic assumption with regard to the quality of feed water is that calcium and magnesium hardness, migratory iron, migratory copper, colloidal silica and other contaminants have been reduced to a minimum, consistent with boiler design and operation parameters.

Once feed water quality has been optimized with regard to soluble and particulate contaminants, the next problem is corrosive gases. Dissolved oxygen and dissolved carbon dioxide are among the principal causes of corrosion in the boiler and pre-boiler systems. The deposition of these metallic oxides in the boiler is frequently more troublesome than the actual damage caused by the corrosion. Deposition is not only harmful in itself, but it offers an opening for further corrosion mechanisms as well.

Contaminant products in the feed water cycle up and concentrate in the boiler. As a result, deposition takes place on internal surfaces, particularly in high heat transfer areas, where it can be least tolerated. Metallic deposits act as insulators, which can cause local overheating and failure. Deposits can also restrict boiler water circulation. Reduced circulation can contribute to overheating, film boiling and accelerated deposition.

The best way to start to control pre-boiler corrosion and ultimate deposition in the boiler is to eliminate the contaminants from the feed water. Consequently, this section deals principally with the removal of oxygen, the impact of trace amounts of contaminants remaining in the feed water, and heat exchange impact.

Feed water is defined as follows:

Feed water (FW) = Makeup water (MW) + Return condensate (RC)

The above equation is a mass balance (pounds or kilograms).

Deaeration (Mechanical and Chemical)

Mechanical and chemical deaeration is an integral part of modern boiler water protection and control. Deaeration, coupled with other aspects of external treatment, provides the best and highest quality feed water for boiler use.

Simply speaking, the purposes of deaeration are:

(1) To remove oxygen, carbon dioxide and other noncondensable gases from feed water

(2) To heat the incoming makeup water and return condensate to an optimum temp

(3) Minimizing solubility of the undesirable gases

(4) Providing the highest temperature water for injection to the boiler

Words and Expressions

1. corrosion [kəˈrəuʒən] *n.* 侵蚀，腐蚀状态
2. dissolved gas 溶解气体
3. carbon dioxide 二氧化碳
4. ammonia [ˈæməunjə] *n.*【化】氨，氨水
5. eliminate [iˈlimineit] *vt.* 排除，消除
6. suction [ˈsʌkʃən] *n.* 吸入，吸力，抽气，抽气机，抽水泵，吸引
7. magnesium [mægˈni:zjəm] *n.*【化】镁
8. deaerate [diˈeiəreit] *vt.* 使除去空气，从（液体）中除去气泡
9. quench water *n.* 急冷水
10. dissolve [diˈzɔlv] *v.* 溶解，解散
11. oxide [ˈɔksaid] *n.*【化】氧化物
12. soluble [ˈsɔljubl] *adj.* 可溶的，可溶解的
13. alkalinity [ˌælkəˈliniti] *n.*【化】碱度
14. hematite [ˈhemətait] *n.* 赤铁矿
15. clarification [ˌklærifiˈkeiʃən] *n.* 澄清，净化
16. filtration [filˈtreiʃən] *n.* 过滤，筛选
17. dealkalization [di:ˌælkəlaiˈzeiʃən] *n.* 脱碱作用
18. demineralization [di:ˌminərəlaiˈzeiʃən] *v.* 去除矿物质
19. impurity [imˈpjuəriti] *n.* 杂质，混杂物，不洁，不纯

Exercises

1. Put the following into Chinese.

(1) water treatment (2) colloidal silica (3) carbon steel

(4) red iron oxide (5) black magnetic oxide

2. Answer the following question, according to text.

(1) What is the most common source of corrosion in boiler systems?

(2) Which aspects are included in Boiler Water Treatment?

3. Translate the paragraph 3, 4, 5, 6 of the text into Chinese.

Unit 12 Air-Conditioning Processes

Maintaining a living space or an industrial facility at the desired temperature and humidity requires some processes called air-conditioning processes. These processes include simple heating (raising the temperature), simple cooling (lowering the temperature), humidifying (adding moisture), and dehumidifying (removing moisture). Sometimes two or more of these process are needed to bring the air to a desired temperature and humidity level.

Various air-conditioning processes are illustrated on the psychrometric chart Fig. 3.18. Notice that simple heating and cooling processes appear as horizontal lines on this chart since the moisture content of the air remains constant (ω = constant) during these processes. Air is commonly heated and humidified in winter and cooled and dehumidified in summer. Notice how these processes appear on the psychrometric chart.

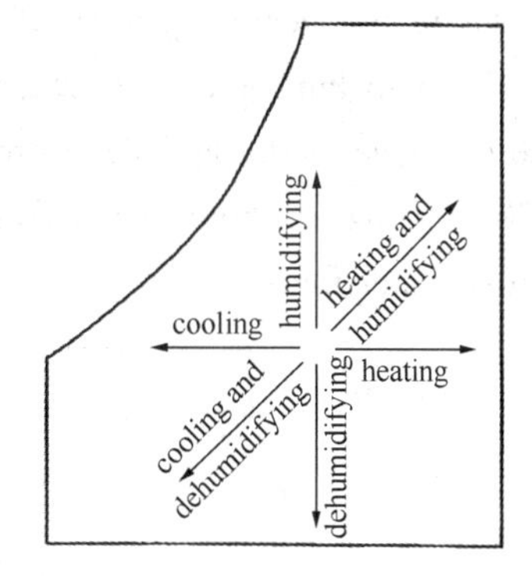

Fig. 3.18 Various air-conditioning processes

Most air-conditioning processes can be modeled as steady-flow processes, and thus the mass balance relation $\dot{m}_{in} = \dot{m}_{out}$ can be expressed for dry air and water as

$$\text{Mass balance for dry air: } \sum \dot{m}_{a,i} = \sum \dot{m}_{a,e} (\text{kg/s}) \quad (3-3)$$

$$\text{Mass balance for water: } \sum \dot{m}_{w,i} = \sum \dot{m}_{w,e} \text{ or } \sum \dot{m}_{a,i}\omega_i = \sum \dot{m}_{a,e}\omega_e \quad (3-4)$$

Where the subscripts i and e denote the inlet and the exit states, respectively. Disregarding the kinetic and potential energy changes, the steady-flow energy balance relation $\dot{E}_{in} = \dot{E}_{out}$ can be expressed in this case as

$$\dot{Q}_{in} + \dot{W}_{in} + \sum \dot{m}_i h_i = \dot{Q}_{out} + \dot{W}_{out} + \sum \dot{m}_e h_e \quad (3-5)$$

The work term usually consists of the fan work input, which is small relative to the other terms in the energy balance relation. Next we examine some commonly encountered processes in air-conditioning.

Simple heating and cooling (ω=constant)

Many residential heating systems consist of a stove, a heat pump, or an electric resistance heater, The air in these systems is heated by circulating it through a duct that contains the tubing for the hot gases or the electric resistance wires, as shown in Fig 3.19. The amount of moisture in the air remains constant during this process since no moisture is added to or removed from the air. That is, the specific humidity of the air remains constant (ω=constant) during a heating (or cooling) process with no humidification or dehumidification. Such a heating process will proceed in the direction of increasing dry-bulb temperature following a line of constant specific humidity on the psychrometric chart, which appears as a horizontal line.

Notice that the relative humidity of air decreases during a heating process even if the specific humidity ω remains constant. This is because the relative humidity is the ratio of the moisture content to the moisture capacity of air at the same temperature, and moisture capacity increases with temperature. Therefore, the relative humidity of heated air may be well below comfortable levels, causing dry skin, respiratory difficulties, and an increase in static electricity.

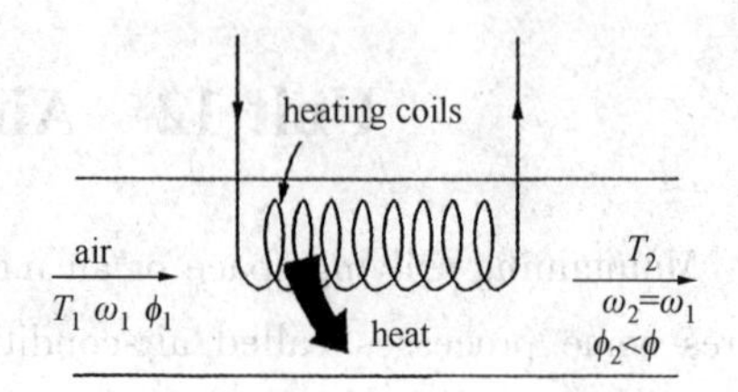

Fig. 3.19 During simple heating, specific humidity remains constant, but relative humiditydecreases

A cooling process at constant specific humidity is similar to the heating process discussed above, except the dry-bulb temperature decreases and the relative humidity increases during such a process, as shown in Fig. 3.20. Cooling can be accomplished by passing the air over some coils through which a refrigerant or chilled water flows.

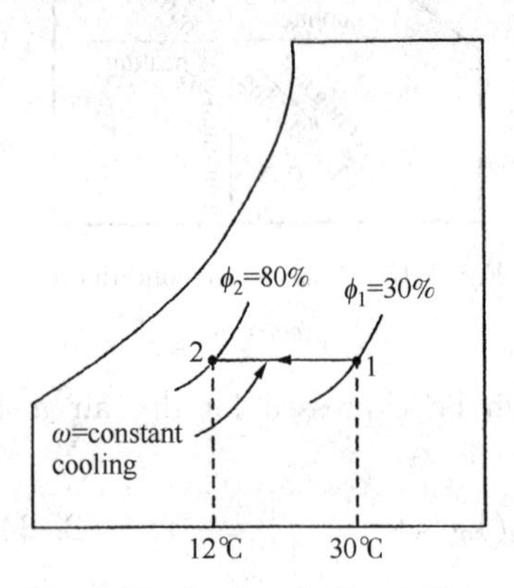

Fig. 3.20 During simple cooling, specific humidity remains constant, but relative humidity increases

The conservation of mass equations for a heating or cooling process that involves no humidification or dehumidification reduce to $\dot{m}_{a1}=\dot{m}_{a2}=\dot{m}_a$, for dry air and $\omega_1=\omega_2$ for water. Neglecting any fan work that may be present, the conservation of energy equation in this case reduces to

$$\dot{Q}=\dot{m}_a(h_2-h_1) \text{ or } q=h_2-h_1$$

Where h_1 and h_2 are enthalpies per unit mass of dry air at the inlet and the exit of the heating or cooling section, respectively.

Heating with Humidification

Problems associated with the low relative humidity resulting from simple heating can be eliminated by humidifying the heated air. This is accomplished by passing the air first through a heating section (process 1-2) and then through a humidifying section (process 2-3), as shown in Fig. 3.21. The location of state 3 depends on how the humidification is accomplished. If steam is introduced in the humidification section, this will result in humidification with additional heating ($T_3>T_2$). If humidification is accomplished by spraying water into the air stream instead, part of the latent heat of vaporization will come from the air, which will result in the cooling of the heated air stream ($T_3<T_2$). Air should be heated to a higher temperature in the heating section in this case to make up for the cooling effect during the humidification process.

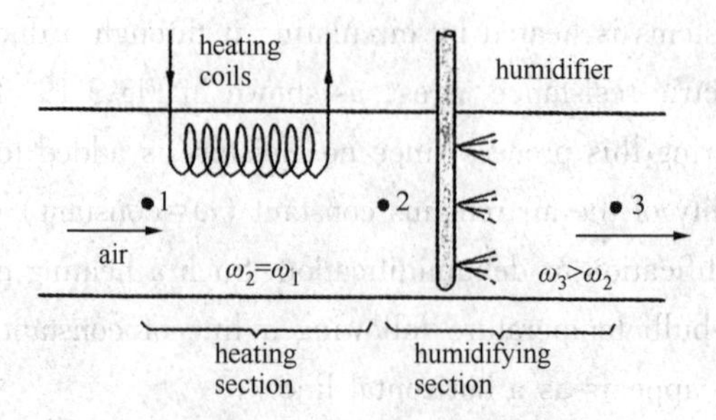

Fig. 3.21 Heating with humidification

Cooling with Dehumidification

The specific humidity of air remains constant during a simple cooling process, but its relative humidity increases. If the relative humidity reaches undesirably high levels, it may be necessary to remove some moisture from the air, that is, to dehumidify it. This requires cooling the air below its dewpoint temperature.

The cooling process with dehumidifying is illustrated schematically and on the psychrometric chart in Fig. 3.22. Hot, moist air enters the cooling section at state 1. As it passes through the cooling coils, its temperature decreases and its relative humidity increases at constant specific humidity. If the cooling section is sufficiently long, air will reach its dew point (state 2, saturated air). Further cooling of air results in the condensation of part of the moisture in the air. Air remains saturated during the entire condensation process, which follows a line of 100 percent relative humidity until the final state (state 3) is reached. The water vapor that condenses out of the air during this process is removed from the cooling section through a separate channel. The condensate is usually assumed to leave the cooling section at T_3.

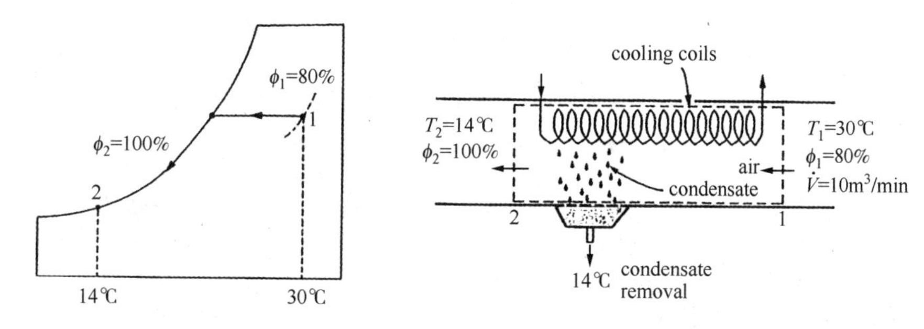

Fig. 3.22 Schematic and psychrometric chart

The cool, saturated air at state 3 is usually routed directly to the room, where it mixes with the room air. In some cases, however, the air at state 3 may be at the right specific humidity but at a very low temperature. In cases, air is passed through a heating section where its temperature is raised to a more comfortable level before it is routed to the room.

Words and Expressions

1. psychrometric chart 湿度计算图，温湿图
2. moisture content 含水量（率）
3. duct [dʌkt] *n.* 管，输送管，排泄管；*vt.* 通过管道输送
4. specific humidity 含湿量
5. humidification [hjuːˌmidifiˈkeiʃən] *n.* 潮湿
6. dehumidification [ˈdiːhjuːˌmidifiˈkeiʃən] *n.* 除去湿气
7. dry-bulb *n.* 干球
8. relative humidity 相对湿度
9. respiratory [risˈpaiərətəri] *adj.* 呼吸的
10. static electricity 静电

11. air stream 气流

12. dew point temperature 露点温度

Exercises

1. Put the following into Chinese.

(1) dry-bulb temperature (2) heating with humidification

(3) cooling with dehumidification (4) saturated air

2. Answer the following question, according to text.

(1) Which processes do air-conditioning processes include?

(2) Why does the relative humidity of air decrease during a simple heating process even if the specific humidity ω remains constant?

3. Translate the paragraph 1, 6, 11 of the text into Chinese.

Unit 13 Shell and Tube Heat Exchanger

A heat exchanger is a device built for efficient heat transfer from one fluid to another, whether the fluids are separated by a solid wall so that they never mix, or the fluids are directly contacted. They are widely used in petroleum refineries, chemical plants, petrochemical plants, natural gas processing, refrigeration, power plants, air conditioning and space heating. One common example of a heat exchanger is the radiator in a car, in which a hot engine-cooling fluid, like antifreeze, transfers heat to air flowing through the radiator.

A shell and tube heat exchanger is a class of heat exchanger designs. It is the most common type of heat exchanger in oil refineries and other large chemical processes, and is suited for higher-pressure applications. As its name implies, this type of heat exchanger consists of a shell (a large vessel) with a bundle of tubes inside it. One fluid runs through the tubes, and the second runs over the tubes (through the shell) to transfer heat between the two fluids. The set of tubes is called a tube bundle, and may be composed by several types of tubes: plain, longitudinally finned, etc.

Theory and Application

Two fluids, of different starting temperatures, flow through the heat exchanger. One flows through the tubes (the tube side) and the other flows outside the tubes but inside the shell (the shell side). Heat is transferred from one fluid to the other through the tube walls, either from tube side to shell side or vice versa. The fluids can be either liquids or gases on either the shell or the tube side. In order to transfer heat efficiently, a large heat transfer area should be used, so there are many tubes. In this way, waste heat can be put to use. This is a great way to conserve energy.

Heat exchangers with only one phase (liquid or gas) on each side can be called one-phase or single-phase heat exchangers. Two-phase heat exchangers can be used to heat a liquid to boil it into a gas (vapor), sometimes called boilers, or cool a vapor to condense it into a liquid (called condensers), with the phase change usually occurring on the shell side. Boilers in steam engine locomotives are typically large, usually cylindrically-shaped shell-and-tube heat exchangers. In large power plants with steam-driven turbines, shell-and-tube condensers are used to condense the exhaust steam exiting the turbine into condensate water which can be recycled back to be turned into steam, possibly into a shell-and-tube type boiler.

Shell and Tube Heat Exchanger Design

There can be many variations on the shell and tube design. Typically, the ends of each tube are connected to plenums (sometimes called water boxes) through holes in tubesheets. The tubes may be straight or bent in the shape of a U, called U-tubes. Fig. 3.23 shows a U-tube heat exchanger.

In nuclear power plants called pressurized water reactors, large heat exchangers called steam generators are two-phase, shell-and-tube heat exchangers which typically have U-tubes. They are used to boil water recycled from a steam turbine condenser into steam to drive the turbine to produce power. Most shell-and-tube heat exchangers are either1, 2, or 4 pass designs on the tube side. This refers to the number of times the fluid in the tubes passes through the fluid in the shell. In a single pass heat exchanger, the fluid goes in one end of each tube and out the other.

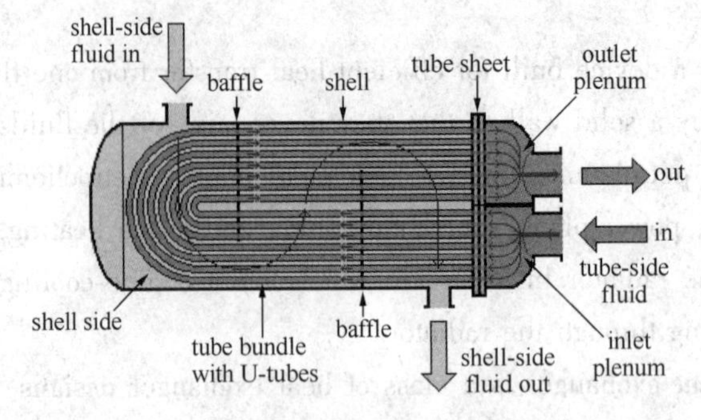

Fig. 3.23 U-tube heat exchanger

Steam turbine condensers in power plants are often 1-pass straight-tube heat exchangers, as shown in Fig. 3.24. Two and four pass designs are common because the fluid can enter and exit on the same side. This makes construction much simpler. Fig.3.25 shows a two pass straight-tube heat exchanger.

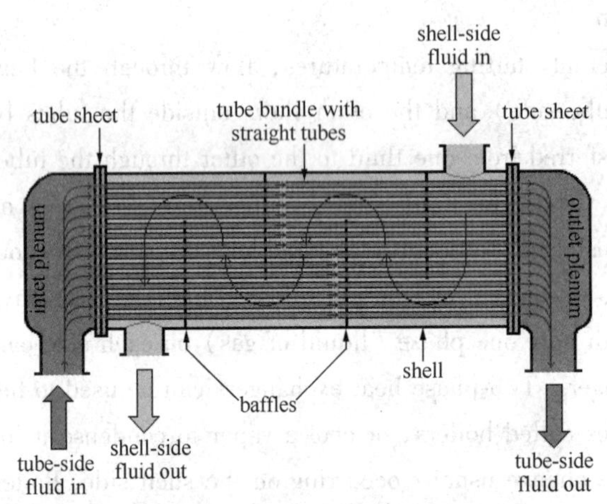

Fig. 3.24 Straight-tube heat exchanger (one pass tube-side)

There are often baffles directing flow through the shell side so the fluid does not take a short cut through the shell side leaving ineffective low flow volumes.

Counter current heat exchangers are most efficient because they allow the highest log mean temperature difference between the hot and cold streams. Many companies however do not use single pass heat exchangers because they can break easily in addition to being more expensive to build. Often multiple heat exchangers can be used to simulate the counter current flow of a single large exchanger.

Selection of Tube Material

To be able to transfer heat well, the tube material should have good thermal conductivity. Because heat is transferred from a hot to a cold side through the tubes, there is a temperature difference through the width of the tubes. Because of the tendency of the tube material to thermally expand differently at various temperatures, thermal stresses occur during operation. This is in addition to any stress from high pressures from the fluids themselves. The tube material also should be com-

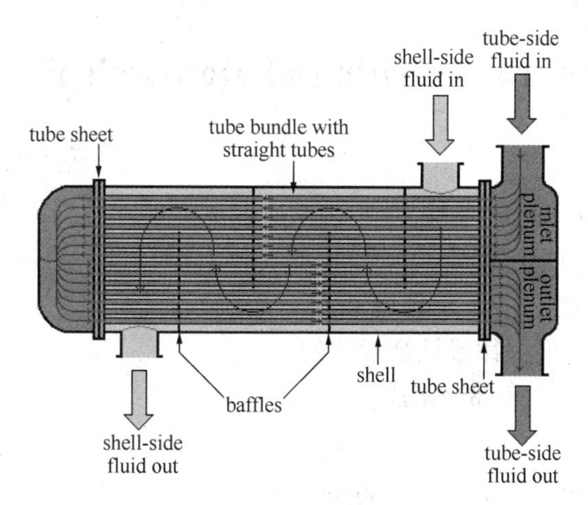

Fig. 3.25 Straight-tube heat exchanger (two pass tube-side)

patible with both the shell and tube side fluids for long periods under the operating conditions (temperatures, pressures, pH, etc.) to minimize deterioration such as corrosion. All of these requirements call for careful selection of strong, thermally-conductive, corrosion-resistant, high quality tube materials, typically metals. Poor choice of tube material could result in a leak through a tube between the shell and tube sides causing fluid cross-contamination and possibly loss of pressure.

Words and Expressions

1. antifreeze ['ænti'fri:z] *n.* 防冻剂
2. tube bundle 管束
3. finned [find] *adj.* 有翼的;带散热片的
4. locomotive [ˌləukə'məutiv] *n.* 机车，火车头
5. exhaust [ig'zɔ:st] *vi.* 排气;*n.* 排气，排气装置
6. baffle ['bæfl] *n.* 折流板
7. deterioration [diˌtiəriə'reiʃən] *n.* 恶化，变坏，消耗，磨损
8. compatible [kəm'pætəbl] *adj.* 谐调的，一致的，兼容的,适应

Exercises

1. Put the following into Chinese.

(1) petroleum refinery (2) chemical plant (3) shell-and-tube heat exchanger

(4) counter current (5) tube side (6) short cut

2. Answer the following question, according to text.

(1) Why counter current heat exchangers are more efficient?

(2) What behaviors should the tube material have?

(3) Weather the boiler is one-phase or two-phase heat exchangers?

(4) How many pass designs on the tube side do most shell-and-tube heat exchangers have?

3. Translate the paragraph 1, 2, 3, 4 of the text into Chinese.

Unit 14　Writing of Research Paper

14.1　总论

科技论文（Research Paper）是记录和讨论研究成果的一种书面表达形式。科研论文无论在词汇、短语、语法的使用还是在篇章结构的安排上都有自己的特点。在体裁上，科研论文作为说明文不同于诗歌、小说，不是用来抒发情感或讲述故事的，而是用说明、记叙等写作手法来提供事实、揭示自然规律或社会规律的。

科技论文一般由以下这几部分组成：

（1）论文题目（Title）

（2）作者姓名（Author's name or Authors' names）

（3）摘要（Abstract）

（4）正文（Body）

（5）参考文献（Bibliography）

14.2　论文题目

论文题目是论文篇首的文字。它是论文内容的高度概括，应充分反映论文的基本内容和特色。论文题目的基本要求是确切、简练、醒目。

论文题目与其他文章的题目一样，是一种特殊的文体。它具有以下一些特征：

（1）在论文题目中，有些冠词是可以省略的。如：

Summary on Development of Switched Reluctance Machine 开关磁阻电机发展综述

这一标题中，在 Summary 前省略了定冠词 The。

（2）论文题目中常用的词类有名词或名词短语。如：

Power Electronics and Harmonic Restraining 电力电子技术与协波抑制

Material Level Measurement and Its Application 料位计及其在火电厂的应用

（3）在论文题目中应避免使用句子，不要使用介词短语或动词不定式短语作论文题目。在必须使用动词的情况下，也可以使用动名词或动词性质的名词，如：

To Apply（或 Applying）the Principle of Phrase Comparison of Fault Component to Transformer Protection 应用故障分量比相原理保护变压器

Application of the Principle of Phrase Comparison of Fault Component to Transformer Protection 故障分量比相原理在变压器保护中应用

（4）另外，在标题中避免使用完整的句子也不是绝对的，有时需要用疑问句作题目，以表示探讨性语气。如：

What Thermodynamics Is About? 热力学的研究对象是什么？

论文题目中常常出现的词汇归纳如下：

summary（综述），development（发展），application（应用），comparison（比较），study（设计），research（研究），experimental research（实验研究），application research（应用研究），design（设计），analysis（分析），method（方法），principle（原则），investigation（调查），discussion（讨论）。

14.3 作者姓名

作者姓名及所在单位编排在题目下面。作者多于一个时，应按字母顺序排，如果论文主要是某一作者完成的，则将该作者排在第一位。

科技论文的作者姓名的英文应写全名，包括名（given name），姓氏（surname）和中间缩写名（middle initial）。此外还要附上作者的专业技术职称、作者的最高学位、工作单位与邮政编码，以便读者与作者联系。如：

Li Junqing　Li Heming

（Department of Electric Power Engineering, North China Electric Power University, Baoding 071003, China）

14.4 摘要

摘要是从文章中摘录的要点，是论文内容的简洁和精确的叙述。摘要内容应包括研究的目的、方法、结果和结论。

摘要要求简练易懂，其原则是尽量以最少的词语表达出论文的基本信息，使读者在最短的时间内可以了解全文的大致信息，从而决定是否需要认真阅读全部论文。因此，写摘要时要正确把握全文的方向，合理地提炼全文，无须进行详细的描写。

为了方便读者，英文摘要应尽可能使用人们较为熟悉的词语，尽量避免使用太冷僻的专业术语，更不要用不正规的词语或行话，也应尽量少用数学表达式或图表。

因为摘要要求写得言简意赅，所以摘要的长度都有字数要求。中等长度的科研论文，其摘要与论文长度之比大约为5%。比较长的论文其摘要与论文之比应低于5%，字数最多不宜超过500个单词。大多数英文摘要的字数应掌握在250个单词以内。短文的摘要更应少于100个单词。

摘要的第一句通常是一带主题性的句子，用以归纳未被论文题目所表达的必要内容。摘要中应尽量用短语代替分句，用单词代替短语。如可以用"to"代替"in order to"，用"seemingly"代替"it would seem that…"。

英文摘要中时态形式出现最多的是一般现在时，其次是现在完成时。英文摘要重点是要叙述科学事实和描述某种过程，而不是强调做这些事情的时间。摘要中常常会表示"报告"什么，"描述"什么，"讨论"什么或结论结果是什么等。表示这些用一般现在时较为恰当。如：The status quo and development tendency of SO_2 emission in coal-burning power plants is described in the paper.

如果是要说明一项科研课题现已取得或已达到的现状时，就宜用完成时。如：The experiment has achieved great success in this respect.

英文摘要中为了突出和强调科学事实、现象和过程的需要，用被动语态要多于或优于用主动语态。如：

The author uses spatial ability test and the Embedded Figure Test to examine the relations among cognitive methods, spatial ability and cartographic achievement.

最好改成：

Spatial ability test and the Embedded Figure Test are used to examine the relations among cognitive methods, spatial ability and cartographic achievement.

文摘中常用到的词汇归纳如下：

be described（描述），be introduced（介绍），be expounded（详加解释），be discussed（讨论），be analyzed（分析），be offered（提供），be pointed out（指出），be inferred（推导），be proposed（提出），be developed（开发），be tested（验证），be testified（验证），be diagnosed（诊断），be used（使用），be built（建立），be adopted（采用），be applied（应用），be investigated（调查），be proved（证明），be compared（比较），be conluded（得出结论），be considered（考虑），be improved（改进），be presented（提出），be studied（研究），be shown（表明），be given（给出），be examined（检查），be reviewed（回顾），be outlined（概括），be revealed（揭示），be put forward（提出）。

在科技论文的英文摘要之后要列出本文的关键词。其目的是为了便于选读和计算机检索存储的需要。关键词必须精选能够代表主要内容的术语。关键词词数规定是3~10个，实际操作中以3~5个词为多见。关键词通常列于简要之后，另起一行开始。关键词或关键词组之间应有2~3个空格或者分号隔开。

下面是一篇论文的题目及摘要，供读者在写作实践中参考。

Summary on Development of Switched Reluctance Machine

Li Junqing　Li Heming

（Department of Electric Power Engineering，North China Electric Power University，Baoding 07100，China）

Abstract

Development，application and structure of the switched reluctancesystem are briefly described. Configuration of stator and rotor，excitation mode and working principle for various switched reluctance machine are expounded. Design methods about the switched reluctance machine are discussed and its advantage，disadvantage and application scope are introduced in detail. Finally，its existing problems and developing trends are analyzed.

14.5　正文

正文是文章的躯干部分，内容应充实，语言表达应准确，简洁，清楚。因此在正文写作中要综合运用科技英语写作中的各种写作方法。

下面就从词、句和段落三个层次来分析一下科技英语写作的特点。

14.5.1　词汇特点

科技论文对词汇的要求主要是准确、简洁。在科技论文写作中，不要使用 may be，might be，could be，possible，possibly，probable，probably 和 likely 等意义模糊的词，也不要使用 think，believe，wish，assume 等带有主观色彩的词汇，而应该用意义清晰，表达准确，不带主观色彩的词汇。

在用词方面，还应该注意简洁，尽量使用简单的词汇，避免使用生僻的、累赘的词汇。也不要使用一些毫无意义的词汇，比如像 in this paper，hopefully，in this connection 等。能用一个词说明的情况就不要用两个或更多的词说明。比如在论文中，用 analyze 就比用 conduct an analysis of 简洁，用 investigate 就比用 make an investigation 简洁。

科技英语作为一种特殊的语域，在用词方面有一些特点。词汇的概念意义是词汇的基本意义，即收录在词典中的意义。如果只知道词汇的基本意义，而不知道该词汇在特定的学术领域中所表达的特殊意义，在论文的写作中就会引起意义的混乱。请看下面的词汇：

trunk 一词的基本意义是“树干，躯干”等，在工程应用中，它的意思是“输电线路的干线”，trunk frame 是“网架”的意思。

field 一词的基本意义是“田野、田地、场地、顿域”等，而它在工程应用中是“现场”的意思，所以 field test 是“现场试验”的意思。transformer 一词的基本意义是 somebody or something that transforms 是“使改变的人或物”的意思，在工程应用中，其意思是“变压器”。再如 conductor 作为普通词汇，意思是“指导者”，而在工程应用中指“导体、导线”；energy 作为普通词汇，意思是“活力”，而在物理学等学科中，是“能量”的意思。power 的基本意义是“能力，权力”，而在科技英语中，意思是“功”。

常见的词汇还有：

a.c.（alternating current 交流电），d.c.（direct current 直流电），max（maximum 最大），min（minimum 最小），s.v.（safety valve 安全阀），T and D（transmission and distribution 输电和配电）。

14.5.2 句子特点

科技论文与普通文体的文章相比，在句子结构的运用方面，也有一些差别。比如被动句的使用等，这些特点是由于科技论文在内容上应力求客观、科学这一现实所决定的。下面就举例说明科技论文在句子这个层次上的特点。

1. 多用长句

在科技论文中要包含的信息量比较大，因此在论文写作中，应多用修饰成分比较多的长句子，避免短句的堆砌。并且少用省略句，甚至关系代词 which，that 都不省略。如：The most economic maintenance strategy during normal operation is the one, where maintenance measures are carried out at the optimum point in time, that is at the point of where damage is just beginning to build up.

2. 科技论文中的时态

科技论文中，句子的主要动词常用一般现在时，因为科技论文要反映的是现阶段所进行的科学研究的情况。一般现在时在科技论文中出现的频率也很高，主要用来反映已取得的成绩或已达到的现状。如：

This has been called the age of electricity, for electricity is now used for such a variety of purposes that it appears to be able to replace all other agencies for doing things. It has caused many changes both inside and outside our homes, but the developments in science to be brought about due to a fuller knowledge of electricity are expected to be even more extensive and fundamental.

我们这个时代被称为是电的时代。因为现今电的用途非常广泛，看来好像能够替代所有的其他手段进行工作。它使我们家庭内外生活发生了许多变化。可以预期，由于对电更加充分认识而引起的科学发展将更加重要。

3. 多用被动语态

在科技论文中，至少有三分之一的动词应用被动结构，因此必须正确理解被动语态的含义，掌握被动语态的用法。

Several approaches to the problem of the driver have been investigated in the past 15 years.

These include changed—particle beams and long and short wavelength and short wavelength and light—ion particle beams are being seriously considered. Although heavy-ion fusion is a promising approach, little progress has been made in the United States, primarily because funding has been limited. Longwavelength lasers which use carbon dioxide were abandoned when it was discovered that wavelength on the order of a micron or more heats the fuel pellet prematurely and caused hydrodynamic instabilities.

4. There be 结构的使用

There be 结构是科技论文写作中的一种常用结构，表示“存在”而不是“占有”的意思。位于句首的 there 是引导词，本身没有意义，句子的主语是 there be 后面的名词或名词短语，be 是系动词，可用各种时态，也可与助动词或情态动词连用。如：

There were some remains in the air tube.

There might be an explosion when heated.

“There+助动词或情态动词+be”是 There be 句型的常见形式。

此外，还有其他一些动词或短语跟在 there 后面也可表示存在的意思，如 exist，remain，stand，live，come 等词。如：

There seemed（appears）to be no doubt about the problem.

There appears to be no impurities in the piece of metal.

5. 以 it 为形式主语的句型

以 it 为形式主语的句型是科技论文中用得非常广泛的句型。如：

It is often necessary to know how much current is flowing in circuit and what voltage. 有必要知道线路中流过的电流多大，电压多高。

不定式的逻辑主语用介词 for 引导，如：

It is possible for free electrons to exist in crystals under the influence of an electric field.

再如：

It is very easy for us to repair such an instrument.

除上面出现的这些形容同外，下面列出的这些形容词也常用这种结构：possible，impossible，advantageous，useful，customary，important，clear，apparent，necessary，easy，difficult，convenient，common，practical，feasible，needless 等。

在这一句型中，在形容词所处的位置上，还可以是名词或过去分词，在不定式所处的位置上还可以是 that 引导的主语从句，it 仍然是形式主语，如：

It is a simple matter with a rectifier to convert A. C into D. C.

It has long been known that the electrical properties of semi—conductors depend very much on their purity.

It was discovered long ago that the neutron had no charge whether positive or negative.

下面归纳了 it 作形式主语时常用的句型：

It is possible that… 有可能

It is impossible that… 不可能

It is important that… 重要的是，是重要的……

It is likely that… 大概，可能……

It is evident that… 是合理的，是适当的

It is appropriate that…是适当的……

It is ture that… 的确，是确定的……

It is questionable that… 成问题的，是有问题的

It is doubtful… 值得怀疑的……

It is notable that… 值得注意的是……

It is noteworthy that… 值得注意的是……

It is conceivable that… 可以想像……

It is worthwhile that… 值得，是值得的……

It is satisfactory that… 令人满意的……

It is good that… 好在，好的是……

It is a fact that… 事实上……

It is no good that… 是无用的……

It is no harm that… 是无害的……

It is a common knowledge that… 众所周知……

It is a good thing that… 好在……

It is a mercy that… 幸而，幸亏……

It is no matter that… 是无关重要的……

It is worth nothing that… 值得注意的是……

It is a common practice that… 通常是……

It is no use that… 无用的……

It is a great pity that… 遗憾的是……

It is the case that… 使人奇怪的是……

It is no wonder that… 难怪，无怪乎……

It is stated that… 据说，一般认为……

It is said that… 据说，有人说……

It is alleged that… 据说，据称……

6. 使用 as 构成的从句及短语

在科技论文中，常常使用到 as 构成的从句及短语。as 在不同的语境中，与不同的词语搭配，意义与用法大不相同。请看例句：

Only in the case of extreme breaking currents and short breaking times are air-blat breakers still used, e.g. as generator breakers with breaking currents in the range of 100-200kA.

这句话中，as 用作介词，意思是“作为，担任”。

Rate of return calculations of this kind are often required in the electric sector, as in other sectors, to justify projects.

as 在这句话中是介词。意思是“像，正如”。

As the circuit-breaker is the last link in the chain of protective equipment the circuit-breaker reliability is of the highest importance for high reliability of the total electrical power system.

as 在这句话中是连词，引导原因状语从句。

The configuration of these early direct digital control (DDC) systems combined analog and digital controls in a true hybrid function as we know that term today.

as 在这句话中引导限定性定语从句, as 在这里相当于 which 或 that。

More formal definitions of real-time are given below, as are the steps of the development of this capability by computer systems and the new advances in information technology.

下面给出的实时的更为正式的定义, 这些定义是通过计算机系统及信息技术的新进展来发展这种能力的手段。

在这句话中, as 作为关系代词引导非限定性定语从句, as 指的是前面整句话的内容

Calculations and nonlinear functions, such as square boot, thermocouple conversions, etc, are easily done. 诸如平方根, 热电偶变换等计算和线性功能易于完成。

such as 意思是"诸如"。

Typically, process-control computer have performed such monitoring and data-logging functions as in-put alarm warning, excursion and transient trending, post-trip review analysis, and sequence-of-events reviewing.

such as 用来表示"列举"的意思。

14.5.3 段落特点

科技论文在体裁上属说明文。说明文的目的是解释事物。解释事物时可以通过定义、分类、举例、比较和对比, 分析其因果关系和提供数据等手段。因此, 在构思段落时, 可以有效地运用这些方式。

科技论文中的段落大多采用主题句—扩展句—结论句的结构, 因为这样的结构反映了科学研究的逻辑严密性。主题句一般位于句首, 它反映了该段文字的中心思想, 也指出了这一段内容的发展方向与方式。扩展句是在主题句之后用以扩展细节, 提供依据的文字。扩展句可以是一句话, 也可以是几句话。结论句与主题句呼应, 是本段文字内容的小节。段落简短时, 常常没有结尾句。下面就举例说明句子发展成一个段落的几种逻辑手段。

1. 定义

定义是揭示概念的内涵或者词语的意义的方法。揭示概念的内涵的定义称为实质定义, 揭示词语的意义的定义称为语词定义。最有代表性的定义是实质定义中的属加种差定义, 即把某一概念包含在它的属概念中, 并揭示它与同一属概念下的其他种概念之间的差别。

The conveyance of electric power from a generation station to consumers known as electric power system. The electric power network is the power of the electric power system but the generators and the consumers, including substations, transmission lines and distribution networks.

在这一例子中, 给"电力网络"这一概念下定义时, 指出"电力网络"的属概念是"电力系统";在"电力系统"这一属概念下, "电力网络"和其他种概念的差别是"它是除去发电机和用户以外的那部分"。

From both an operation and functional point of view the power network can be divided into several substructures based upon operating voltage level. Highest on the voltage scale is the transmission or the power grid.

这一段实际是给输电网 (power grid) 或电力网络 (power network) 下了个定义, 指出"输电网"的属概念是"电力网络", 在"电力网络"这一属概念下, 输电网和其他种概念的差别是"highest on voltage" (电压最高)。

A substation is an assemblage of equipment for the purpose of switching of changing or regulating the voltage of electricity.

这一段是给变电站下了个定义。在下定义时，常常要用到以下这些词和短语：be，be defined as，be known as，be called，mean，refer to，involve，deal with，be concerned with.

2. 分类

分类是揭示概念的外延的表达方式，把具有共同特性的有关事物分成属类。它的目的是系统化，使有共同特征的事物系统的组合在一起。在科技论文写作中，分类是非常有效的一种组织文章的策略。请看下面的例子：

Heat is transferred by three basic modes. In two of these modes the heat is transferred through material, whereas in the third mode it is transmitted through empty space and vacuum as well as through certain materials transparent to thermal radiation. The physical mechanisms and laws for the heat transfer are different for each of these basic modes, which are designated as conduction and radiation.

这一段文字对热量的传递方式进行了分类。

Main classifications of the substations are as follow：Step up substation，primary grid substation，secondary substation and distribution.

这一段对变电站进行了分类。

在对事物进行分类时常要用到这些词语：class，group，category，kind，sort，be classified into，be classed into，be divided into，fall into，group into，be made up of.

3. 举例

举例就是使用例子来说明要点。它是最常见，也是最有效的一种说明方式。好的例子能深入浅出地使抽象的事物具体，同时还能提高读者的兴趣以增加论文的说服力。请看下面两例：

Topography often determines the type of dam. For example，a narrow V shaped channel may dictate an arch dam. The topography indicates surface characteristics of the valley and the relation of the contours to the various requirement of the structure. Soundness of the rock surface must be included in the topographic study.

这段文字中的主题句是 Topography often determines the type of dam. 而后就是用举例子的方式扩展文章的。

These various forms of an element are called"isotope". Most elements have more than one form，or isotope. Oxygen has three. Uranium has two isotopes. The more common one has 92 protons and 146 neutrons，with an atomic weight of 238. The less common isotope of uranium also has 92 protons but only 143 neutrons，and an atomic weight of 235. The two isotopes are usually referred to as "U-238" and "U-235". About one out of every 140 uranium atoms is U-235，and the rest are U-238.

这段文章的主题句是 These various forms of an element are called "isotope".（一种元素的这些不同形式被称为"同位素"）。即，这段文字主要是说明"同位素"这一概念。接着文章用氧和铀这两种元素作例子进一步说明同位素这一概念。

在举例时，常常要用到这些词语：like，an example of，such as，be illustrated by，for example，as illustrated by，for instance。

4. 比较与对比

比较是确定事物同异关系的思维过程和方法，即根据一定的标准把彼此有某种联系的事物加以对照，从而确定其相同与相异之处，以便对事物进一步初步的分类。对比则是把两种不同的事物作对照，互相比较。请看例子：

The most important way in which gases differ from liquid and solid is that they can easily be compressed, or squeezed to a smaller volume.

这段文字中比较了气体与固体和液体的不同之处。

The hydroelectric source has the advantage that it is immediately (within second at least) available. Whereas thermal sources can meet demand at a much slower rate, hydroelectric sources have the disadvantage that their use is constrained by navigation requirements and actual or predicted rainfall.

这里把水利发电和火力发电进行了比较。

类比是一种特殊比较形式。类比不是为了表示同一类别的两个事物的相同之处，而是为了表示不同类别的两个事物的相同之处。类比是用 Y 来解释 X 的。如：

Just as water will always try to find its lowest level, heat will always flow a body a high temperature to one at a low temperature, and it is impossible for heat from a cool body to flow of its own accord into a hot body (making it hotter still.) This is called the second law of thermodynamics.

这里将热的流动与水的流动作了个类比，用水的现象来解释热的现象。

在对事物进行比较和对比时，常用到以下词语：comparison, contrast, both, similar, difference, different, similarities, like, while, although, alike, same, on the other hand, in a similar way, on the contrary, however, comparing to, instead, unlike。

5. 因果分析

因果分析法用于阐述和分析某事件产生的原因和结果。某一事件产生的原因是什么？造成了什么结果？这都需要用因果分析法来说明。请看例子：

The human-machine interface may be the most important aspect of the entire ISCS. It is through this interface that the substation operator must control and monitor the entire substation. The data must be presented to the operator in a clear and precise manner. There must be no possibility of ambiguity or error, because of the critical of operator actions on substation equipment as well as personnel safety concerns.

这一段说明了人机界面为什么是整个集成变电站综合自动化系统 ISCS 的最重要的方面。第一句是结果，接下来解释了原因。

在解释原因和结果时，常用到这样的词语：as, because, since, accordinly, consequently, hence, therefore, thus, so, as a result, thus, so, because of, account for.

14.6 参考文献

在科技论文的写作中，参考文献占有重要地位。参考文献不仅是为方便读者查阅，它本身也反映了作者的治学作风是否严谨，对他人的学术成果是否尊重。因此，科技工作者在撰写学术论著时，一定要认真地处理好参考文献。参考文献一词在英文中有几种不同的表达方式，如：references, bibliography, reading list 等。

学术期刊和专著的编排方式，国际上通用的作法是这样的：

期刊：作者 + 文章题目 + 期刊名 + 卷号 + 起止页码。

专著：作者 + 年份 + 出版单位 + 起止页码。

参 考 文 献

1 Holman J P. Heat Transfer (Ninth Edition). McGraw-Hill, 2002
2 Mills A F. Heat Transfer (Second Edition). Prentice Hall. Upper Saddle River, NJ, 1999
3 KATSUIKO OGATA. Modern Control Engineering (Fourth Edition). Prentice Hall. Copyright, 2002
4 Cengel Y A, Boles M A. Thermodynamics: An Engineering Approach (Fourth Edition). McGraw-Hill, 2002
5 李瑞扬. 热能与动力工程专业外语. 哈尔滨:哈尔滨工业大学出版社, 2004
6 阎维平,冯跃武. 热能与动力工程专业英语. 北京:中国电力出版社, 2006
7 沈国富,唐俭等. 新编电力英语教程. 北京:电子工业出版社, 2004
8 靳智平. 电厂汽轮机原理及系统. 北京:中国电力出版社, 2004
9 程明一,阎洪环等. 热力发电厂. 北京:中国电力出版社, 1998
10 金滔,陈国邦. 制冷与低温专业英语. 北京:中国电力出版社, 2005
11 杨小灿. 现代英汉制冷与空调词汇. 北京:国防工业出版社, 2001
12 Stambuleanu A. Flame Combustion Processes in Industry. England: Abacus Press, 1976
13 Junkai Feng. Science and Technology of Industrial and Utility Applications. Coal Combustion. H.P.C, 1988